# Servicing Radio, Hi-Fi and TV Equipment

*Books on Radio and Television by Gordon J. King*

Audio Equipment Tests
Audio Handbook
Beginner's Guide to Colour Television
Beginner's Guide to Radio
Beginner's Guide to Television
Colour Television Servicing
Newnes Colour Television Servicing Manual
Practical Aerial Handbook
Servicing with the Oscilloscope
Television Servicing Handbook

# Servicing Radio, Hi-Fi and TV Equipment

Gordon J. King

**Heinemann: London**

William Heinemann Ltd
10 Upper Grosvenor Street, London W1X 9PA

LONDON   MELBOURNE   JOHANNESBURG   AUCKLAND

First published by George Newnes Ltd 1966
Reprinted 1968
Second edition 1973
Reprinted 1975, 1978
Third edition (retitled) 1982
Reprinted 1983
First published by William Heinemann Ltd 1987

© Gordon J. King 1982

**British Library Cataloguing in Publication Data**
King, Gordon J.
  Servicing radio, hi-fi and TV equipment. – 3rd ed.
  1. Transistors – Handbooks, manuals, etc.
  I. Title II. Rapid servicing of transistor equipment
  621.3815′28      TK7871.9

ISBN 0 434 91071 6

Printed in Great Britain at the University Press, Cambridge

# Preface

The aim in this book has been towards fault diagnosis, and to embrace the maximum amount of equipment chapters have been written to deal with specific sections focused upon the two essential parts of any equipment, namely, amplification and signal generation. Various types of oscillators and amplifiers are dealt with, and finally included is a chapter on fault-finding in the ordinary transistor portable, which is now so extremely popular. The transistor hi-fi amplifier has also been considered, and examples of commercial circuits are given.

Early chapters introduce the basic semiconductor and transistor and reveal their mode of operation. Chapters are included to show how transistors are set up in circuit, how they are biased and tested, and one complete chapter deals with signal conditions and tests.

Each 'fault-finding' chapter concludes with a Fault Diagnosis Summary Chart. There are a total of five such charts, and each one gives probable causes of fault conditions and items to check to aid diagnosis. A complete chart is given for fault diagnosing in the transistor portable.

It is assumed that the reader possesses a basic understanding of electronic principles and circuit techniques. It is not supposed, however, that the reader is an electronics engineer, for the text is directed to the service technician, to the student starting a career in electronics and to the enthusiastic amateur, many thousands of whom find enjoyment in experimenting with transistor circuits and devices.

At this point I have great pleasure in giving my sincere thanks to the many individuals and firms who have helped to make this book possible. In particular I should like to thank Mullard Limited for the information that they have given to me for publication.

Special thanks also go to Pye Limited for the use of equipment for experiment, test and photographing, and for permission to publish details of their equipment. Also to the many firms who have kindly permitted me to reproduce circuit diagrams, circuit sections or technical details of their products. I am also grateful to the firms who supplied photographs for publication.

I was glad of the opportunity to update this little book and make it into a third edition which, with its new title, it is hoped will continue to be as popular in the future as it has most certainly been in the past. Since the book first appeared in 1966 solid state devices have ousted valves in almost all spheres of electronics, so the aim set then to present the book 'as though thermionic valves had never existed' is even more applicable today.

In this Third Edition I have taken the opportunity to delete out-dated information and add new, but the original plan has been preserved since much of the information previously presented is just as important to the service technican as ever it was.

Brixham, Devon. *Gordon J. King*

# 1 Transistor fundamentals

It is possible to trace faults in transistor circuits without any knowledge of how the device works. The exercise is made that much easier, however, with at least a basic understanding of what goes on inside the device. One can become highly proficient in the art of diagnosing troubles in transistor equipment without the need to delve deeply into the physical science of crystals and semiconductors. But a scientific approach to these subjects is required by those whose job is to develop transistors and design circuits.

In this book we are concerned with the practical aspects of transistors, other semiconductor devices and their circuits, not with design and development. For those whose wish it is to delve deeply designwise many excellent books are available.

This book aims to bridge the gap between the theoretical and the practical aspects of semiconductors with the main emphasis on rapid fault finding.

Early literature dealing with the transistor tended to apply comparison between the transistor and the thermionic valve and between their circuits. Such comparisons can be useful in one way, though confusing in another, especially to young and student technicians starting from square one who have no practical knowledge of the valve behind them. To the old hand brought up on valves the comparison technique has much to commend it. In this book we shall adopt the comparison technique as little as possible. We shall approach transistors as though valves barely existed!

Having now set the stage, let us get on with the act ...

## Semiconductors

A transistor is created within a crystal. The two chief crystals used are germanium and silicon. Pure crystal is a very good electrical insulator. However, to work with a transistor the crystal must be arranged to conduct electricity in a rather special way. This is done by introducing to the pure crystal a form of impurity. The crystal thus starts off very pure and then is modified by the controlled addition of the impurities. This changes the pure crystal from a very good insulator to a *semiconductor*.

A semiconductor can be considered as a material which is neither a good insulator nor a good conductor but something between the two.

Any conductor of electricity must have within its makeup *current carriers* which are available for the conduction of electricity. Copper is well known for its good conductivity, and like all other materials, is composed of atoms which feature a central nucleus around which electrons orbit. The charge of the nucleus is positive while that of the electrons is negative. The atom is electrically balanced since the positive charge equals the negative charges of the electrons. The electrons are bound in their orbits to the nucleus. The electrons in orbit close to the nucleus cannot easily be moved from the orbit, but those in an outer orbit are less tightly bound. In copper and other good conductors there is an abundance of outer orbit electrons which are continuously shifting from the outer orbit of one atom to that of another. These are sometimes called *free* or *mobile electrons*. They are current carriers.

An electrical potential applied across a conductor (such as the application of a battery across a length of copper wire) causes the free electrons to move along the conductor from negative to positive. The battery can be considered as a 'pump' which takes in the electrons and then pushes them back to the negative end of the conductor again.

An orderly movement of electrons constitutes a flow of electricity. Thus, electricity flows through a conductor when it is connected across a battery. Since electrons are negatively charged they are referred to as *negative current carriers*. It may be as well to note here that although electrons move from negative towards positive (electron flow), an electric current is said to flow from positive towards negative (conventional current flow). Before the electron theory was fully understood it was published that electricity had a flow from positive to negative, and this idea stuck even when it was fully realised that electrons actually flow from negative to positive.

Unlike charges attract, as is well known, and the electrons being the mobile carriers thus move towards a positive charge.

Now, the more free electrons possessed by a material, the greater is its electrical conductivity. A material with no free electrons is a perfect insulator. A pure crystal has no free electrons.

Crystal used in the creation of a transistor is given current carriers by the addition of so-called *impurity*. Arsenic is an impurity which is often introduced to germanium to produce a semiconductor. When arsenic is added complicated things take place within the crystalline structure, but basically we can look at it like this. Germanium can be considered as having four orbiting electrons tightly bonded to each of its atom nuclei. These are not mobile and cannot serve as current carriers. However, arsenic has five orbiting electrons to each of its atom nuclei. The effect of the addition is that an atom of arsenic replaces an atom of germanium in the crystal. This causes four of the arsenic electrons to pair with the four electrons of an adjacent germanium atom. One electron per arsenic atom thus remains unpaired. This is free and represents a mobile current carrier.

The processed germanium thereby turns from an insulator into a semiconductor of specific characteristics whose current carriers are electrons. This type of semiconductor is called 'n-type' (*n* for *n*egative electron).

## n- and p-type semiconductors

From the foregoing discussion the reason for denoting the semiconductor n-type will not be appreciated, since it will be considered that all conduction must be by way of negative electrons. This is really true, but in semiconductor science another type of current carrier is created, as we shall see.

Another type of semiconductor is produced by the addition of an impurity to the basic crystal which has only three orbiting electrons to each of its atoms. Such an impurity is indium.

Now, when this is added to germanium its three electrons pair with three electrons of an adjacent germanium atom, thereby leaving one germanium electron per indium atom without an electron partner. This electron deficiency is termed a *hole*. This implies that there is, in fact, a vacancy for an electron in the crystalline material. Holes thus exist throughout the crystal structure and conduction is said to occur as the result of a movement of holes.

A battery connected across a length of this kind of semiconductor produces a displacement of the indium electrons so that they move from hole to hole. That is, an electron moves from one hole into an adjacent hole. This leaves a hole which is then filled by an electron of a near indium atom behind it. While the electrons are moving from negative to positive, the holes are effectively moving from positive to negative, as bascially illustrated in Fig. 1.1.

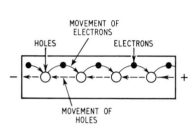

Fig. 1.1. Electron vacancies or '*holes*' exist in p-type semiconductor material. An electric potential applied across this type of semiconductor causes electrons to move from hole to hole in a negative to positive direction and the holes themselves to move in a positive to negative direction. The direction of movement is shown by the arrow heads. The holes are considered as the current carriers. N-type semiconductor has spare electrons as current carriers

Since the holes are electron deficiencies they are, in fact, positive charges, and for this reason are termed *positive current carriers*. This type of semiconductor is called 'p-type' (*p* for *p*ositive).

## Semiconductor junction

A piece of crystal semiconductor by itself is of little use. In a transistor both p- and n-type semiconductors are employed, arranged in the form of junctions.

Before we can understand just how the transistor works, it is necessary to investigate the semiconductor junctions in a little detail. A basic junction is formed when a p-type semiconductor is brought into contact with an n-type semiconductor. In practice, the union is not achieved by bringing together the two types of processed crystal. There is basically a single piece of crystal which is processed in such a manner that one side or one end of it develops into an n-type semiconductor while the other side or other end develops into a p-type semiconductor. A p-n junction then occurs within the crystal between the two types of material.

Now, when the crystal has been processed in this way a so-called *potential barrier* is created at the junction. The effect is that some of the electrons from the n-type semiconductor diffuse across the junction and neutralise some of the holes in the p-type semiconductor. This means that in the proximity of the junction the n-type semiconductor assumes a positive charge due to the loss of some of its electrons while the p-type semiconductor assumes a negative charge due to the loss of some of its holes. These positive and negative charges across the junction give rise to a potential difference which in semiconductor parlance is termed a potential barrier. This is illustrated in Fig. 1.2.

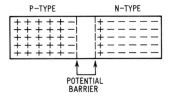

Fig. 1.2. Due to diffusion of holes from the p-type semiconductor and electrons from the n-type a potential barrier is formed across a semiconductor p-n junction, negative on the 'p' side and positive on the 'n' side

The potential of the barrier cannot be measured by putting a voltmeter across a junction diode, for example, as the potential exists only at the boundaries of the junction. Nevertheless, its effect can be demonstrated.

If we apply a battery across the ends of the p- and n-type semiconductors with a current meter in series we find that with the battery connected one way round we read very little current while with the battery connected the opposite way the current value is very much higher. We get a diode effect, in fact.

Knowing about the potential barrier across the actual junction, we can see from Fig. 1.3 why the diode effect occurs. At (a) the positive

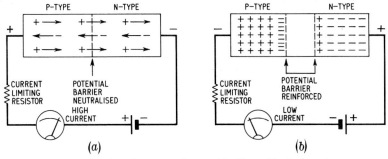

Fig. 1.3. By the connection of an external potential with positive to the p-type semiconductor and negative to the n-type (a), the potential barrier is neutralised allowing free interchange of holes and electrons across the junction. Under this condition the junction is biased for forward conduction and a high current is indicated on the meter. The resistor in series is purely for current limiting.

With the battery polarity reversed, as at (b), the potential barrier is reinforced and forward conduction is impossible. The extremely small current flowing under this condition is caused by thermally generated 'minority carriers' and is called 'leakage current' of the battery is connected to the p-type semiconductor and the negative to the n-type. By looking at Fig. 1.2 it will be seen that the battery applied this way round puts an opposing potential across the potential barrier. This has the effect of neutralising the barrier. Electrons from the n-type semiconductor then flow into the p-type and holes from the p-type flow into the n-type.

### Forward conduction

A free interchange of positive and negative current carriers thus takes place as shown in Fig. 1.3(a). An easy path for electric conduction exists and the meter reads high current. With the potential applied this way round the junction is said to be biased for *forward conduction*.

### Reverse conduction

With the battery connected the opposite way round, as shown in Fig. 1.3(b), the potential barrier is reinforced and through current in the normal way is impossible. Under this condition the junction is said to be biased for *reverse conduction*. An extremely low current may be indicated on the meter. This is called *leakage current* and results from conduction by *minority carriers*, about which more will be said later.

### Diode effect

The semiconductor junction thus exhibits a rectifier or diode effect. It passes current easily in one direction only. Here, then, we have the *junction diode*.

Similar effects arise when the junction is formed of a semiconductor and a metal, such as the cats-whisker of a crystal detector or of a *point-contact diode*. Diffusion, in fact, can take place between two metals in which there is a difference between the make-up of the current carriers in each.

### Summary

To sum up. Although pure crystal is a very good insulator, it is made semiconducting by the addition of an impurity which releases into the crystalline structure current carriers. Depending upon the impurity added, the majority current carriers are either electrons or holes. Semiconductor in which electrons are the current carriers is n-type while that in which holes are the carriers is p-type. Holes and electrons have the same charge magnitude but opposite sign – electrons negative and holes positive.

Owing to their opposing signs, electrons and holes are attracted towards each other, and a process which is fundamental to the operation of a semiconductor is that of an electron filling a hole. This is called *recombination*, and when it happens neither the electron nor the hole can serve to carry current.

This business of holes is not always easy to grasp, particularly when one has been brought up on electrons as current carriers. As long as we realise that the impurity added to produce p-type semiconductor results in fewer mobile electrons than available spaces (holes) which they could occupy we are on firm ground. In this type of semiconductor electrons from neighbouring atoms move in and out of the surplus holes. New holes are thus being continuously created in place of the original ones. In this way the holes 'wander' round the semiconductor in a random manner, or they flow through in a controlled manner when a potential is applied across the semiconductor (Fig. 1.1).

The atomic impurities added to the basic crystal to produce p- and n-type semiconductors are sometimes called *acceptors* and *donors* respectively, i.e. those that create holes which accept electrons from neighbouring atoms and those that donate extra electrons to the crystal structure.

To understand exactly how the electrons and holes are created as required in the semiconductor we should have to make a very close study of the structure of solids in terms of atoms, molecules, bonding, crystal lattices and so forth. While such study would, indeed, be highly instructive, it is not essential for the readers of this book.

When the junction between the two types of semiconductor is processed during manufacture, holes and electrons diffuse across it only until a state of equilibrium exists. The diffusion then ceases. It occurs again, however, when the junction is biased for *forward conduction*, as we have seen.

## Minority carriers

Before we go on there is one more point to consider. That is the small current that flows through a junction diode when it is biased for reverse conduction. This happens because of minority carriers. A minority carrier is a mobile electron in p-type semiconductor and a hole in n-type semiconductor. Minority carriers arise mainly because of the impossibility of achieving the perfect in practice.

It will be understood that their presence encourages conduction in the reverse sense across a junction. Normally, there are so few of them compared with the majority carriers of opposite sign that the reverse conduction is extremely small. However, minority carriers do tend to multiply as the temperature of the junction is increased, due to thermal agitation effects which release a greater number of mobile carriers than normal from the atoms into the conduction stream.

Leakage currents due to minority carriers are extremely important factors in transistors as we shall see in later chapters, and precautions must be taken against them. Just to illustrate the point at this stage, the reverse current given by the set-up in Fig. 1.3(b) would be greatly increased if the junction were held in front of a heat source.

The leakage current is also increased by light falling upon the junction. The physical reason for this is complex, but basically, the light produces hole-electron pairs at both sides of the junction. The effect of the potential barrier is that the holes are deflected one way and the electrons the other, giving rise to an effective increase in leakage current. The current above that of normal leakage produced by the light is called *photoelectric current*.

Ordinary transistors are protected against light effects by the use of a type of case which excludes light; modern encapsulating techniques have led to the use of plastic cases as distinct from the early metal ones. The phototransistor actually exploits the light effect, the photoelectric current at a p-n junction being amplified by the transistor.

Thus, we have seen that a semiconductor junction is affected both by heat and light.

## Diode characteristics

We must now investigate the characteristics of a junction diode. Let us suppose that we set up the circuit in Fig. 1.3(a) so that the voltage applied is continuously variable from zero up to a reasonably high value. We would find that the current would rise gradually at first and then more steeply. If we plotted the forward current against the forward voltage we should get a characteristic as shown in Fig. 1.4(a).

The curve shows that there is not an immediate current flow. This is because of the barrier potential, and it is only when the barrier potential has been countered by the applied voltage that current begins to flow, at point A on the characteristic. The voltage value between points B-C divided by the current value (in amperes) between points C-D gives the slope resistance of the diode. Silicon

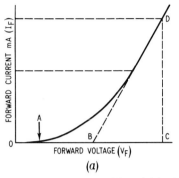 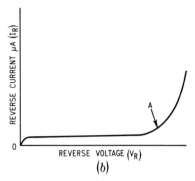

Fig. 1.4. Diode characteritiscs: (a) the forward characteristic and (b) the reverse characteristic

crystal power diodes have a slope resistance of only a fraction of an ohm, which is the cause of their great efficiency, while small signal diodes have a slope resistance somewhat greater. The curve at (a) represents the *forward characteristic*.

Now, if we set up the circuit in Fig. 1.3(b) so that the applied voltage is continuously variable from zero upwards, current and voltage plotting as before would produce a curve rather like that in Fig. 1.4(b). This is the *reverse characteristic*.

The current flowing this time is extremely small over a normal voltage range (it being the leakage current). At a certain highish voltage point, depending upon the type of diode, there is a sudden increase in current, at point A on the characteristic. This is called the *turnover voltage* or *avalanche breakdown voltage*. It is this which sets a limit to the *inverse rating* of the diode. It occurs as the result of a sudden multiplication of minority carriers, and the value, as would now be expected, is influenced by junction temperature.

## Zener effect

The voltage at which a semiconductor junction breaks down when biased in the reverse direction is also called the *zener voltage*, after the name of Dr. Carl Zener, the scientist who commenced early explorations into the reverse breakdown mechanism of junction diodes.

For normal applications semiconductor diodes are not subject to an inverse voltage that approaches the zener or breakdown voltage, but there are times when such a diode is purposely back-biased to the extent of the breakdown point. Now, when a diode breaks down in this manner there is a sudden and substantial increase in reverse current flow as indicated at point A in Fig. 1.4(b). This change from normal leakage current to *zener current*, as it is sometimes called, is so abrupt that the reverse of zener voltage holds substantially constant for relatively large changes in zener current.

This characteristic is exploited for voltage regulators, voltage reference devices, as surge and overload protectors, as signal limiters and clippers and a host of other things.

This means that if a diode is back-biased from a supply circuit through a series resistor, the voltage applied to a load circuit from across the diode will hold substantially constant in spite of changes in load current. Moreover, the zener voltage will hold constant even though the voltage of the input supply may vary somewhat. Here, then, is a simple voltage regulator using the zener effect.

What really happens in the circuit is that if the load takes an increasing current, the current in the series resistor will tend to increase likewise. But this is prevented from happening by the diode passing a smaller zener current, thereby holding the total current in the series resistor constant. The zener voltage across the diode thus remains constant, meaning also that the voltage across the load is held constant.

There are many configurations of zener diode regulation systems, but all of them are based on the foregoing action.

It is interesting to note that all junction and semiconductor diodes exhibit this zener effect, but not all diodes are suitable for voltage regulation applications. Diodes in which the zener effect can be ulitised fully are called *zener diodes*, as might be expected. They are also called *reference diodes* and *voltage regulator diodes*, among other things.

Zener diode application can be seen in Fig. 7.9 in Chapter 7 (page 188). Note how the zener diode is distinguished from an ordinary diode by the little tail on the cathode part of the symbol. In the circuit mentioned, R1 is the series resistor that contributes to the control action, the zener voltage here being used to regulate the voltage at the emitter of the control transistor Tr2.

### p-n junction characteristic curve

A diode characteristic has the two curves integrated into one graph, as shown in Fig. 1.5. Points to note are how the forward and reverse characteristics are affected by temperature, especially the reverse

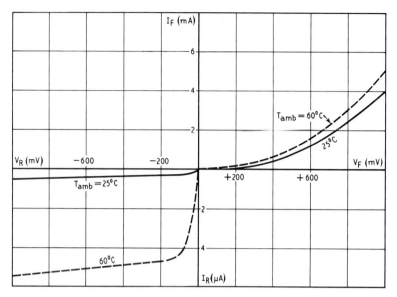

Fig. 1.5. Complete diode characteristic. Note how the characteristics are affected by temperature, especially the reverse characteristic which signifies leakage current

characteristic, and that the forward current is given in milliamperes (mA) while the reverse current is given in microamperes (μA), these being thousandths and millionths of an ampere respectively.

## Shorthand

The shorthand used so far is $I$ for current and $V$ for volts. Thus, for forward current and voltage we have $I_F$ and $V_F$ while for reverse

Fig. 1.6. Conventional diode symbol. The arrow head is the anode and the heavy line the cathode. The symbol for the zener diode can be seen in Fig. 1.33

current and voltage we have $I_R$ and $V_R$ respectively. Subscripts are commonly used in semiconductor parlance to indicate across what or in which the current $I$ or voltage $V$ refers.

The conventional diode symbol is shown in Fig. 1.6. The heavy line is often called the 'cathode' and the arrow head the 'anode'. These are colour-coded red and black respectively or may be marked with plus and minus signs. The arrow head corresponds to the p-type semiconductor and the heavy line to the n-type. Forward current direction is signified by the arrow head. Forward current flows when this is made positive with respect to the cathode.

## The transistor

In one way a transistor behaves like two junction diodes with a common electrode. The common electrode is called the *base*, which may be either n- or p-type semiconductor. If it is n-type it is sandwiched between two p-type layers while if it is p-type it is

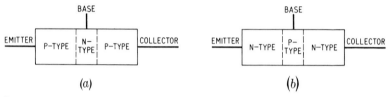

Fig. 1.7. Geometric representation of the junction transistor: (a) p-n-p type and (b) n-p-n type. Note the two 'diode' junctions

sandwiched between two n-type layers. Either way two semiconductor junctions are formed, p-n-p or n-p-n, with the middle semiconductor being common to both junctions. Transistors are known as p-n-p or n-p-n type because of this. The semiconductors either side of the centre bases are called *collector* and *emitter*. The basic geometrical conception of the transistor is shown in Fig. 1.7 while the symbol is shown in Fig. 1.8, p-n-p and n-p-n at (a) and (b) respectively in each drawing.

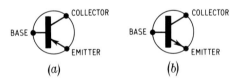

Fig. 1.8. Transistor symbols: (a) p-n-p type and (b) n-p-n type. Note the difference in direction of the arrow head of the emitter between the two types

The junction on the emitter side is called the *emitter junction* and that on the collector side of *collector or junction*. In some transistors both junctions are almost identical while in others there is a considerable difference between them.

In action, the emitter junction is usually biased for forward conduction and the collector junction for reverse conduction. The simplest way of achieving this is shown in Fig. 1.9, at (a) for a p-n-p transistor and at (b) for an n-p-n transistor. Note the reversal of voltages at the emitter and collector between the two circuits.

If we put a current meter in series with the base circuit we would expect to get a reading of forward current. We should, in fact, get such a reading, but it would not be a very big one on a small transistor, just a matter of microamperes. In a practical circuit this might be produced by about 200 mV between the emitter and base (i.e., $V_{eb}$). The current is limited by the elements – too great a current would ruin the transistor.

Now, we would not normally expect to obtain a significant current reading in the collector circuit owing to the fact that the collector/base junction is reverse-biased. We may be surprised, therefore, to find that a milliampere or so of current is flowing.

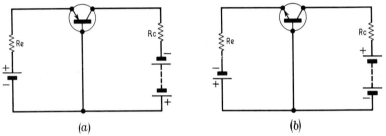

Fig. 1.9. Emitter and collector junction biasing with two batteries: (a) with a p-n-p transistor and (b) with an n-p-n transistor. $R_e$ and $R_c$ are respectively emitter and collector resistors. Both (a) and (b) are common-base circuits

After ensuring that the polarity of the collector battery is correct, we would ultimately discover that the relatively high collector current flows only when forward current is flowing in the base/emitter junction. We would find on further investigation that the current flowing in the collector circuit is merely the collector/base leakage current due to the reverse-biased collector/base junction when the battery forward biasing the base/emitter junction is disconnected. In short, we would have discovered the transistor effect!

That is, collector current flows only when the base/emitter junction is biased for forward conduction. If we continued testing along these lines, we would discover that the amount of collector current is related (up to a certain point) to the amount of forward current flowing in the base/emitter junction. We would also discover that the emitter current is equal to the collector current *plus* the forward current in the base/emitter junction, which is fairly obvious when one comes to think about it. However, before we become too deeply involved here, let us get some idea of why the transistor effect occurs.

Let us take the case of a p-n-p transistor. The emitter and collector are of p-type semiconductor and the base of n-type. Set up in a circuit such as in Fig. 1.9(a), the emitter acts as a source of positive holes. This is because it is of p-type semiconductor.

The holes flow into the n-type base because this is negative with respect to the emitter for forward conduction in the emitter junction. Remembering that unlike signs attract, we appreciate why the holes, which are positive charges, flow into the negatively charged base. Almost all the holes diffuse through the base and enter the region of the collector junction. The holes are attracted to the collector because this is also negative (with respect to the base), the flow into it thereby producing collector current.

A relatively high collector current thus occurs in spite of the collector junction being biased for reverse conduction.

The amount of collector current will, of course, depend upon the amount of holes diffusing through the base from the emitter. As more holes are produced by increasing the forward current in the emitter junction, it follows that this action will produce an increase in collector current. A few of the emitter holes combine with the base electrons, but the resulting loss of charge is made good by the flow of base current.

Clearly, then, as the emitter-base forward current is increased, so is the current in the emitter-to-collector circuit. The current flowing in the base circuit is referred to as the 'base current'. This is very much smaller than the emitter current which is made up of the base current *plus* the collector current.

So much, then, for the p-n-p transistor, but what about the n-p-n counterpart? The action of this is very similar to that already described. The emitter this time acts as a source of electrons because it is of n-type semiconductor. Thus, when the emitter is biased in the forward direction a relatively large current of majority carriers (electrons) will cross the emitter-base junction. Electrons pass into the p-type base and in this semiconductor represent minority carriers. They diffuse into the base and eventually reach the base-collector junction. Since the base is this time biased positively with respect to the emitter it accepts the electrons and collector current results as before.

The electrons reaching the base-collector junction can be considered as being 'swept' by the barrier potential, which exists at that junction due to its reverse bias, into the n-type collector region, thereby providing a collector current governed by the flow of minority carriers across the forward-biased emitter-base junction. This will be better understood by referring to Fig. 1.3(b) which shows the positive barrier potential on the n-type semiconductor when this is connected to a positive source, as is the collector of the n-p-n transistor (see (b) in Fig. 1.9). This positive barrier potential, of course, attracts the electrons which diffuse through the base. As before, some of the electrons diffusing through the base tend to combine with the holes in the p-type semiconductor. This results in a loss of charge, but is made good by the continued flow of base current.

## Basic transistor types

For efficient transistor action, the base section has to be very thin. One type of junction transistor consists roughly of a thin bar of semiconductor crystal whose bulk is n- or p-type. In the centre of the same bar is a processed, thin layer of p- or n-type semiconductor, this being part of the same single crystal.

Fig. 1.10. Photomicrograph of a section of a high-frequency alloy diffused transistor etched to show the various junctions (Mullard Ltd)

For very high frequency working, transistors with extremely thin base sections are demanded, and techniques are used to reduce the transit time of the current carriers operating between the emitter and collector. The *alloy diffused* type of transistor (Fig. 1.10) is common for v.h.f. work. Here the emitter and base are attached to the same side of the *collector* semiconductor. The assembly is processed by carefully controlled r.f. heating, and by the use of pellets for the emitter and base which are caused to diffuse into the p-type collector.

The base pellet is of n-type semiconductor and the emitter pellet of p-type, thereby giving the p-n-p transistor. The emitter pellet also contains some n-type impurity, so that when heated and diffusion occurs the n-type impurity penetrates the crystal more deeply than the p-type impurity. This results in a very thin, graded n-type base layer between the p-type collector and the p-type emitter. Grading of this nature produces a so-called 'drift field' which accelerates the holes from the emitter into the barrier potential attraction of the collector. In that way the carrier transit time is reduced and the transistor is able to function at very-high and ultra-high frequencies.

Another arrangement is seen in the *mesa* transistor, so called because of its construction, mesa meaning a table-shaped hill. The collector semiconductor crystal forms the 'table' upon which the transistor is built, and during manufacture a double diffusion process is employed. The first heating 'diffuses' the base-collector junction while the second results in the formation of the base-emitter junction.

The assembly is finally etched and metallised contact areas are applied to the base and emitter for the connection of fine, gold wires. The base is only about 0.0025 mm thick and is sandwiched between the emitter and collector.

Also used for v.h.f. and u.h.f. applications is the silicon *epitaxial planar* transistor. Planar implies that the transistor is formed upon a flat surface, this, in fact, being the collector semiconductor crystal. The n- or p-type impurity is diffused into the collector to produce a base region and a similar process is afterwards adopted to form the emitter region. Control over the location of the base and emitter regions is achieved by first oxidising the surface of the collector crystal wafer, producing a coating of silicon oxide which prevents diffusion. The base and emitter regions are diffused through 'windows' etched in the oxidised surface, the surface being reoxidised after each operation. In common with other diffused transistors, the base layer can be carefully controlled and hence made very thin. This means that the time taken for holes or electrons to pass through the base region is much reduced so that very high frequency of operation is possible – up to several thousand MHz. The oxidised surface, protecting the collector base junction, results in very low and stable collector leakage current, giving improved performance. To reduce the internal resistance of the collector region, and hence the voltage drop across it, a compound wafer consisting of a higher resistance collector region nearer the junction on a low-resistance substrate is used. The higher resistance region is a thin film grown on to the single crystal substrate: the orientation of the crystal structure from the low to the higher resistance region has to be maintained, hence the name epitaxial – in the same axis.

Power transistors need extra good emitter efficiency and this is arranged by making the conductivity of the p-type emitter region greater than that of the n-type base region. Power transistors are also designed (see Fig. 1.11) for use with heat sinks, their purpose being to 'drain' away the heat developed in the transistor by conduction. Such transistors are clamped in good thermal contact with a bulk of metal or fin for cooling.

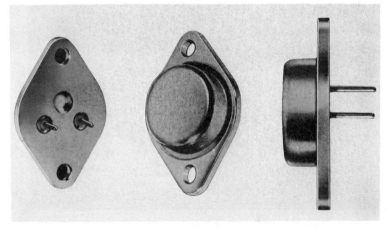

Fig. 1.11. Samples of power transistors by Mullard. Note the heavy case which is clamped on to a bulk of metal to conduct heat away from the junctions. The case is in electrical connection with the collector

Most small signal transistors now employ the crystal silicon. The semiconductor properties of processed crystals of both types are similar, but silicon works at higher temperatures than germanium, it has a lower leakage current and is less temperature sensitive than germanium. Gain is thus maintained at very low temperatures.

For the sake of continuity, it should be mentioned that very early transistors were of the point-contact type. They featured a single crystal of n-type semiconductor with two fine wire points (cat's-whiskers) pressing on to it a few thousandths of an inch apart. These points gave the emitter and collector and the crystal the base. As this type of transistor is no longer used, it will not be referred to again.

To sum up, then, on the transistor action. We have seen that when current is varied in the low impedance forward direction of the

18  Transistor fundamentals

emitter-base junction, current also varies in the high impedance collector circuit. The low impedance exists because of the forward biasing of the emitter-base junction while the high impedance collector characteristic exists because the collector-base junction is

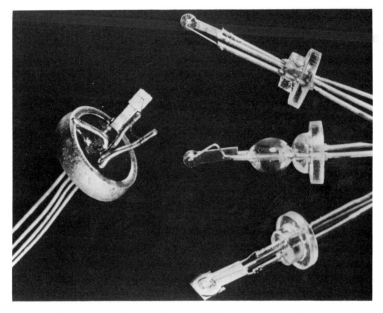

Fig. 1.12. This photograph, approximately 3.5 times actual size, shows three Mullard transistors and a crystal diode before encapsulation. The types are left AF114, top right OC45, middle right OA1; (diode) and lower right OC81 (Mullard Ltd)

biased for reverse conduction. This impedance difference allows the transistor to be used as a voltage amplifier, as we shall see.

### Transistor circuit modes

Fig. 1.9 shows circuits of the *common-base* mode (or earthed base). This means that the base is common to both the input and output circuits. We have seen that if we injected current into the emitter-base junction, in series with the emitter resistor $R_e$, we should get a change in current in the collector resistor $R_c$, relative to base. Thus, the base is common to input current and output current.

The other two modes, *common-emitter* and *common-collector*, are shown in Fig. 1.13 (a) and (b) respectively. It will be seen that p-n-p transistors are featured and that this time only a single battery is employed for biasing both junctions. The emitter-base junction gets its low voltage bias from the junction of the potential divider, R1 R2, which is connected across the single battery. This makes the base of the transistor a little negative, depending upon the ratio of R1 and R2, relative to the emitter which is connected to battery positive. The collector, of course, is negative relative to both base and emitter.

Exactly the same techiques can be adopted with n-p-n transistors, but the battery is connected the opposite way round to forward-bias the emitter junction and reverse-bias the collector junction.

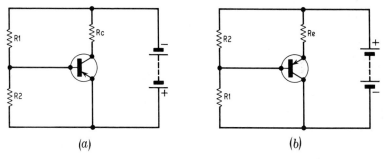

(a)    (b)

Fig. 1.13. The junction of R1, R2 is negative with respect to the emitter of a value depending upon the ratio of the resistors. The base is thus made negative with respect to the emitter for the required degree of forward conduction. (a) is a common-emitter circuit and (b) a common-collector circuit. The resistor $R_e$ in (b) is used both as the load and for d.c. stabilisation, as explained in the text

The potential-divider arrangement for biasing the base will be found in the majority of circuits, as it possesses advantages over other arrangements. An arrangement which may be found in simple circuits is shown in Fig. 1.14. Here base-emitter current is passed through $R_b$ from the collector side of $R_c$.

## Thermal runaway

We have seen that a junction biased for reverse conduction will pass a little leakage current due to minority carriers. If the junction is allowed to warm up, as the collector junction may under operating conditions or if the circuit is subjected to high ambient temperatures, the leakage current will rise owing to a multiplication of minority carriers passing the potential barrier. This effect will cause a rise in junction current and a consequent further rise in temperature. This will produce even more current and more heat, until eventually the dissipation of the junction will be exceeded and the transistor destroyed. This is called *thermal runaway*.

Transistor circuits embody protection against this. In Fig. 1.14, for instance, a rise in collector current due to minority carrier conduction

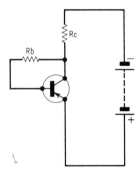

Fig. 1.14. A method of obtaining base bias from a single battery by the use of resistor $R_b$. This is adjusted in value to provide the requisite base forward current. A degree of d.c. stabilisation is provided by returning the top of $R_b$ to the collector instead of direct to supply negative

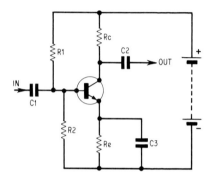

Fig. 1.15. The classic common-emitter circuit in which $R_e$ is used to assist with stabilisation, as explained in the text. The input is applied to the base and the output is taken from the collector, relative to emitter in each case. Note the n-p-n transistor and supply polarity

will increase the volts drop across $R_c$ and make the collector less negative. This will reflect back to the base as a reduction in negative bias and will thus reduce the normal collector current.

In the circuit in Fig. 1.13(b) a similar action takes place due to the emitter resistor, $R_c$. Since a rise in collector current would also give a similar rise in emitter current, the volts drop across $R_e$ would increase should $I_c$ tend to rise abnormally. The effect is to make the emitter go more negative with respect to the base. As this is the same as the base going less negative with respect to the emitter, the forward current in the emitter junction is pulled back and the collector current is reduced.

The circuit in Fig. 1.13(a) has no protection at all and would not normally be used in this form. An emitter resistor to protect the transistor (n-p-n in this case) would be included, as shown in Fig. 1.15. To prevent this resistor affecting the signal it is usually bypassed with a capacitor, C3 in the circuit. As this is a common-emitter circuit, the signal is applied to the base and taken from the collector, via coupling capacitors C1 and C2 respectively, both input and output signal being a common to the emitter.

The protection devices described serve to stabilise the d.c. conditions. Variations of these will be found in practical circuits, and in some circuits d.c. feedback is applied, via d.c. coupling, from one stage back to another.

## Static characteristics

We now know sufficient about transistors and their circuits to appreciate how they work. Before we get down to detail, however, it is useful to see how the current in the base-emitter junction rises as the base-emitter voltage is increased. This is shown for a small signal transistor by the curve in Fig. 1.16. This is called the *input characteristic* in the common-emitter mode. Since the input is to the base, the base current ($I_b$) is plotted against the base voltage ($V_b$), it being understood that the base-emitter junction is concerned.

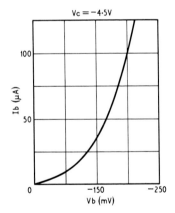
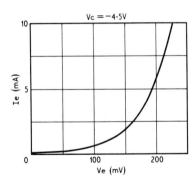

Fig. 1.16. This input characteristic shows how the base current $I_b$ rises with increase in base voltage $V_b$ in the common-emitter mode

Fig. 1.17. Input characteristic in the common-base mode. Here emitter current $I_e$ is plotted against emitter voltage $V_e$

The curve is very non-linear at the start and then straightens up at high $V_b$. The base current would probably be set to about 60 µA by the application of a base voltage in the order of −175 mV, the straight part of the curve then being utilised. The curve goes to show the small voltages and currents in the base. It also reveals the relatively low input impedance, typically 500 ohms to 1000 ohms, but, of course, the actual value depends upon the base current flowing. At very low currents the impedance is higher than it is at higher currents.

In the common-base mode the input characteristic is obtained by plotting the emitter current ($I_e$) against the emitter voltage ($V_e$), as shown in Fig. 1.17. Here the input impedance ($V_e/I_e$ at any selected point on the curve) is even lower, typically 50 ohms to 100 ohms. Again, the curve is non-linear at the start.

These non-linearities are countered by feeding in the signal current through a source impedance which is high compared with the input resistance of the transistor and by operating the transistor on a linear part of the curve. Since the transistor requires a current input, however, the signal could vary the input impedance. It is for this reason that a constant current input is needed which is not affected by changing impedance.

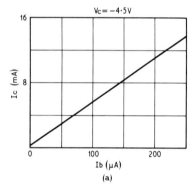 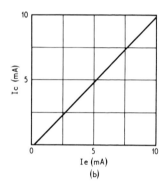

Fig. 1.18. Transfer characteristics: (a) in the common-emitter mode and (b) in the common-base mode. The ratio $I_c/I_b$ gives the current gain in the common-emitter mode and the ratio $I_c/I_e$ gives the current gain in the common-base mode

A curve which shows how the collector current ($I_c$) changes with changes of either $I_b$ or $I_e$ is called a *transfer characteristic*. $I_c$ against $I_b$ is plotted for the common-emitter mode and $I_c$ against $I_e$ for the common-base mode. Such characteristics are shown at (a) and (b) respectively in Fig. 1.18. These curves are plotted with a fixed $V_c$, the value of which is usually indicated upon them.

### Current gain

Examining curve (a) we see that about 14 mA of $I_c$ flows for a $I_b$ of 250 µA. We thus have a ratio of 14 000 µA to 250 µA. This works out to a value of 56, which is the *current gain* of the transistor. The current gain at any other point on the curve can be found in a similar manner, but since this characteristic is linear the current gain will be pretty well constant at all points.

Common emitter current gain is signified by *beta* ($\beta$), *alpha dash* ($\alpha'$) or $h_{fe}$. Such gain at zero frequency (i.e., d.c.) is signified by a zero subscript on the *alpha dash* symbol, giving *alpha nought dash* ($\alpha_0'$) and by $h_{FE}$ (i.e., the use of the subscript capitals 'FE'). The current gain at signal frequency may not necessarily be the same as that at d.c. owing to various effects that the transistor can have upon the signal, depending upon its frequency.

We should be able quickly to recognise the various symbols given to current gain and other transistor parameters.

Curve (b) shows that the current gain for the common-base mode ($\alpha$) is slightly less than unity. This is because the $I_c$ must always be less than the $I_e$ by an amount equal to $I_b$. A typical value is 0.98.

## Power gain

Power gains are possible in all modes, of course, since the power change at the input (either at the base or emitter) is always less than the power change at the output (either at the collector or emitter). This is because the emitter-base junction is forward-biased with a fraction of a volt and a fraction of a milliampere (low impedance) while the collector-base junction is reverse-biased (high impedance) with several volts and yet the collector current is in the order of milliamperes due to the transistor effect.

Power gains are in the order of 10 000 times in the common-emitter mode, 1 000 times in the common-base mode and 200 times in the common-collector mode.

## Voltage gain

Voltage gains are possible because a high impedance load may be connected in the collector circuit. The collector is then fed through the load from a relatively high voltage source. A relatively large voltage change will thus occur across the collector load due to a very small change in voltage at the base or emitter input. A small transistor is capable of stepping up a signal voltage by as much as 10 000 times or more.

## Input and output impedance

Apart from having different power gains, the different transistor modes or circuit configurations give a range of input and output impedances. Typical values are as follows: common-emitter mode 2 000 ohms input 50 000 ohms output, common-base mode 150 ohms input 200 000 ohms output and common-collector mode 40 000 ohms input and 500 ohms output. The various modes are for this reason found as impedance matching devices as well as amplifiers.

A common-emitter circuit has the input applied at the base and taken from the collector, a common-base circuit in at the emitter and out at the collector and a common-collector circuit in at the base and out at the emitter.

## Output characteristics

The high collector impedance is revealed in the output characteristics. These are shown in Fig. 1.19 at (a) for a common-emitter circuit and at (b) for a common-base circuit. In both cases the $I_c$ is plotted against the $V_c$ at various values of $I_b$ at (a) and $I_e$ at (b).

The very high output impedance of the common-base circuit is shown by the collector current being almost the same at all collector voltages above the *knee voltage*. The knee voltage is that voltage where the characteristic straightens out. Up to the knee voltage $I_c$ rises rapidly with increase in $V_c$.

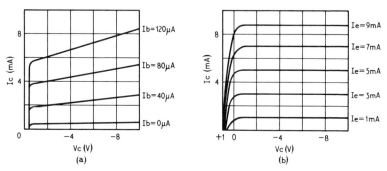

Fig. 1.19. Output characteristics in the common-emitter mode (a) and the common-base mode (b). The high output impedance is signified by the fact that a relatively large swing in collector volts $V_c$ produces only a small change in collector current $I_c$

We have so far seen most of the characteristics in which we may be interested during exercises of fault tracing in transistor circuits. We may not use them very often, but it is as well to know about them, just in case.

## Leakage currents

All simple types of transistor tester measure either d.c. or a.c. current gain (or both) *and* leakage current, usually that in the collector junction.

It will be recalled that leakage current is that current which flows across a reverse-biased junction due to thermally generated minority carriers. Good transistors at normal ambient temperature should have low leakage.

The leakage current in the collector junction, between collector and base, with the emitter floating is called $I_{cbo}$. $I$ stands for current and cb between collector and base. The o implies that the remaining electrode is floating or open-circuit.

By a similar token, the leakage current in the emitter junction, with the collector floating, is $I_{ebo}$. These leakages are usually very small, of the order of 3 to 10 µA, depending upon the temperature.

Some transistor testers give a test for leakage current between the emitter and collector. That is, $I_{ceo}$, sometimes referred to as $I'_{co}$. This parameter is no longer given by transistor manufacturers because it is not a true measure of leakage. The fundamental leakage measured in this instance is $I_{cbo}$. However, the leakage must flow through the base-emitter junction due to the nature of the test. It thus initiates a $I_c$ which is beta times as great, due to the normal transistor action. Thus $I_{ceo}$ is not a true leakage, but is a leakage plus gain function of the the transistor. On one transistor $I_{ceo}$ may be as low as 50 µA while on another transistor of the same type it could be as high as 400 µA. Its measurement thus tends only to confuse.

## Coupled pairs

Circuits containing pairs of transistors will frequently be encountered. These provide specific conditions, such as high input impedance and low output impedance (buffer amplifier) or very high current gain ('super alpha' or Darlington pair). As d.c. coupling is used between the two transistors of the pair, the two should be regarded as a single 'circuit block' when servicing. A useful buffer

26  Transistor fundamentals

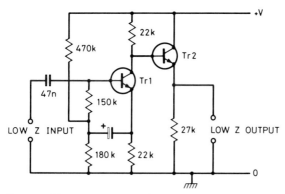

Fig. 1.20. This two-transistor pair exhibits a high input impedance and low output impedance (about 3.6 MΩ and 250 Ω respectively). The first transistor is in common-emitter configuration and the second in common-emitter (emitter-follower) configuration. Biasing of the second transistor is established by the $V_c$ of the first. The large amount of feedback round Tr1 sets the gain which, in this circuit, is unity, thereby making it useful as a buffer amplifier with impedance matching capability. Suitable transistors would be BC148 or equivalent

amplifier by this type is shown in Fig. 1.20. Details of other circuits are given in my companion book *Radio Circuits Explained*.

## Other semiconductor devices

Although the simple bipolar transistor and junction diode are the most frequently used devices at the time of writing, hosts of different devices have appeared over the last few years, some of which are related to the basic junction diode and transistor actions and others which have different principles of operation.

As a start, let us have a look at some of the developments of the diode which in some guise or other will be found in the latest equipment under service.

## Capacitor-diode

A capacitor-diode or varicap, as it is sometimes called, is basically a junction diode in reverse conduction. The symbol is shown is Fig. 1.21. The effect is that the depletion region at the junction, where the 'potential barrier' develops, acts as a kind of dielectric between two 'plates' formed by the n-type and p-type semiconductors. Since the depletion region widens as the reverse bias is increased this is tantamount to the two plates of a capacitor being moved away from each other. Thus the capacitance value falls as the reverse bias is increased.

The leakage current in the reverse bias mode tends to detract from the goodness factor ($Q$ value) of the capacitor, but the varicap diodes which are developed specially to exploit this principle are remarkably 'efficient' and are, in fact, used at u.h.f. in television tuners and at v.h.f. both in television tuners and f.m. receivers. The capacitive component of the diode is arranged to shunt the inductor of the tuned circuit such that as the reverse bias is altered so also is the tuned

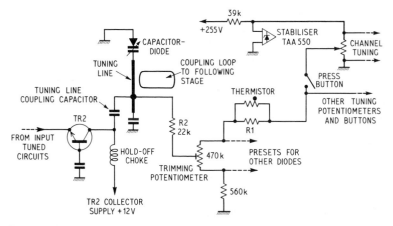

Fig. 1.21. Capacitor-diode tuning of one stage of a u.h.f. television tuner

frequency. The bias is usually regulated by a potentiometer geared to the tuning mechanism or to the push-buttons and sometimes the frequency synthesising of a television front-end, and the control potential is fed to all the diodes of the variable tuned circuits of the r.f., mixer and local oscillator stages so that the single potentiometer changes the frequency of all the tuned circuits together and in step, as required by the application.

In some designs front-end alignment or 'trimming' is facilitated by a small preset potentiometer being arranged to adjust separately the bias to each capacitor-diode. The usual 'padding' capacitors might also be required for optimum tracking over the band, while the inductors themselves might incorporate the usual screw cores for low-frequency r.f. alignment. Capacitor-diodes are available in matched sets.

Since frequency drift would occur from a drift in bias potential, the bias source is often a simple transistor or integrated circuit regulator or zener diode stabilising circuit (page 188). Moreover, automatic frequency correction is facilitated by the use of capacitor-diode tuning, since it is then only a simple matter of applying the control potential from the f.m. detector, in the case of an f.m. radio receiver, in series with the biasing potential to counter frequency drift of the local oscillator or other tuned circuits. Sometimes the control potential is applied only to the local oscillator capacitor-diode, since this is the most critical tuned circuit of the front-end so far as frequency stability is concerned.

In television receivers the i.f. channel feeds a separate phase discriminator circuit the d.c. output of which swings either positively or negatively depending on the direction of mistuning, and it is this output which constitutes the control potential, correcting the tuning error.

The circuit in Fig. 1.21 shows the application of a capacitor-diode in the tuner of a colour television receiver. Here the diode tunes the tuning line in the collector circuit of one of the transistors. With the push-button depressed, the d.c. circuit is from the 225 V input, via the stabiliser and channel tuning potentiometer, the thermistor and padding resistor R1, the trimmer potentiometer, R2, the tuning line itself and back to chassis. Thus the capacitor-diode is reverse-biased and the amount of bias is adjustable by the channel tuning potentiometer and, to a smaller degree, by the trimming potentiometer. Notice the symbol used for the capacitor-diode. The thermistor provides temperature compensation.

## Tunnel diode

The tunnel diode is a development of the ordinary junction diode, but instead of the forward current rising steadily with increasing forward voltage, it reaches a peak and then commences to *fall* with further increase in forward voltage, thereby exhibiting a negative resistance characteristic. After falling into a valley, the current commences to rise again and the normal diode characteristic is resumed, as shown in Fig. 1.22. It is instructive to compare this with the ordinary diode forward characteristic in Fig. 1.4(a).

The negative resistance effect can be exploited in oscillator, switching (in logic circuits) and even amplifying circuits.

The unusual characteristic results from the valence electrons of the semiconductor atoms near the junction 'tunnelling' across the junction from the p-type region to the n-type region. Similar 'tunnelling' occurs in some devices under conditions of small reverse bias, which makes the diode highly conductive for all reverse biases, which of

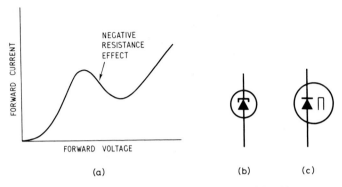

Fig. 1.22. Characteristic of tunnel diode (a) and symbols (b) or (c)

course differs from the normal diode reverse characteristic (Fig. 1.4(b)).

The effects result from heavy doping of the semiconductor materials used for the diodes. Forward conduction thus commences at a much lower value of forward voltage than required by the ordinary junction diode to combat the potential barrier or 'contact potential'. Various crystalline materials are used for fabricating tunnel diodes; one being gallium arsenide which has particular advantages when the device is used as a microwave oscillator.

## Other diodes

Other diodes mostly of use for generating microwave energy include avalanche and impatt diodes, impatt standing for *imp*act *a*valanche and *t*ransit *t*ime, the Read diode, rather like the tunnel diode, which oscillates in accord with the dimensions of an associated cavity, the Trappett diode suitable for relatively low-frequency pulse work, the Schottky barrier diode, based in one fabrication on a junction formed by silicon and a metal, and the step-recovery diode which has some of the characteristics of a capacitor-diode but in the more specialised form for frequency multiplication – varactor diode. The p-i-n diode, which acts as a sort of variable resistor governed by signal input, is now bring used in some TV and f.m. front-ends in the form of a variable attenuator to minimise overloading, thereby increasing the front-end dynamic range.

Mixer-diodes are often used at frequencies of 1 000 MHz or more, where this sort of frequency is reduced to a lower i.f. value. Such diodes are often designed to fit into a coaxial-cable-terminated waveguide.

Light-emitting diodes (LEDs), made of gallium arsenide or indium phosphide crystals, have an electrical behaviour similar to that of ordinary semiconductor diodes, but the energy liberated during forward conduction manifests as visible light radiation from the junction. LEDs are commonly used as low-consumption indicator lights and signal-level indicators, such as on cassette tape decks, where their swift-response function can be exploited. The light emission can also be modulated by varying the forward current. With a photo-diode as the receiver, short-path communication becomes possible. Infra-red LEDs are used in radio and TV remote-control units, with an infra-red photo-diode at the receiver responding to the digitally encoded pulses, which operate the various functions.

## Field effect transistors

While the transistor so far discussed relies on two types of current carrier (electrons and holes), the field effect transistor (f.e.t.) uses only one type. For these reasons the f.e.t. is sometimes called unipolar and the other species bipolar. F.E.T.s using electron carriers are n-channel ( for *n*egative electrons) and those using hole carriers are p-channel (*p* for *p*ositive holes).

Instead of the output current being controlled by the base/emitter current, f.e.t. control is by a voltage at the input electrode called the *gate* and roughly equivalent to the base of a bipolar transistor. The electrode from which the carriers flow is appropriately called the *source*, which is basically equivalent to the emitter of a bipolar transistor, and that to which the carriers flow is, again very appropriately, called the *drain*, basically equivalent to the collector of the bipolar transistor.

F.E.T.s have different symbols from bipolar transistors, and Fig. 1.23 shows the symbols of the more common varieties. Before we can fully appreciate these, however, there are one or two features of the f.e.t. which will have to be explored. Firstly, because the f.e.t. is 'voltage controlled', as distinct from the forward current control in the base/emitter junction of a bipolar transistor, it exhibits a much higher input impedance than the bipolar transistor.

The high input impedance results from the gate electrode being 'isolated' from the rest of the device either by a diode section in reverse conduction or by a layer of actual 'insulation'. The nature of the 'isolation' endows the device with a specific name. An f.e.t. in

Semiconductors 31

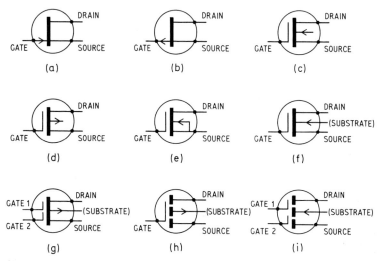

Fig. 1.23. F.E.T. symbols. See text for full description

which the gate operates via a reverse-biased diode section is called a *junction gate* f.e.t., while that which incorporates 'insulation' is called an *insulated gate* f.e.t. As metal oxide commonly forms the 'insulation' the letters MOS may prefix the name, standing for *metal oxide semiconductor*.

Secondly, the nature of the device determines the style of gate biasing required. One type may need to be reverse-biased to secure the correct operating point, rather like a thermionic valve because output current flows even when the gate is at the same potential as the source. This is called a *depletion* type. Another behaves more like a bipolar transistor in terms of biasing; that is, it needs to be forward biased to obtain drain current and the correct operating point. This is called an *enhancement* type.

The f.e.t. symbol signifies the type of device, and it is noteworthy that some f.e.t.s incorporate two gate electrodes. In practice, a control potential – such as that for automatic gain control – is sometimes fed to one gate, while the actual signal for amplication is fed to the other. When such an f.e.t. is used as a mixer, the local oscillator signal may be fed to one gate and the incoming signal to the other. Some devices also have a connection to the substrate.

We can now return to Fig. 1.23, where at (a) is given the symbol of the junction gate n-channel f.e.t., at (b) the same but p-channel, at (c) the insulated gate depletion n-channel f.e.t., at (d) the same but p-channel, at (e) the insulated gate depletion n-channel (with the substrate internally connected to the source) f.e.t., at (f) the same but with the substrate connection brought out, at (g) insulated two-gate depletion n-channel (with the substrate connection brought out) f.e.t., at (h) the insulated gate enhancement p-channel (with the substrate connection brought out) f.e.t. and at (i) the insulated two-gate enhancement n-channel (with the substrate connection brought out) f.e.t. Although these are the symbols recommended by the British Standards Institution, variations are often found in circuit diagrams.

It will be seen that each symbol includes an arrow head, and it is the direction in which this is pointing which shows the polarity of the device. When the arrow is pointing into the symbol it is n-channel and when it is pointing out it is p-channel. This has some correspondence to p-n-p and n-p-n bipolar transistors, as can be seen by reference to Fig. 1.8, the arrow on the emitter pointing inwards on the former and outwards on the latter. p-channel f.e.t.s generally require a negative drain and n-channel a positive drain. Fig. 1.24 shows the circuit of a

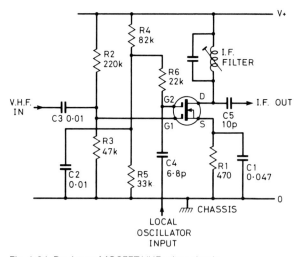

Fig. 1.24. Dual-gate MOSFET VHF mixer circuit

v.h.f. (f.m.) mixer using a dual-gate MOSFET, where gate 1 receives the input v.h.f. signal and gate 2 the local oscillator signal. Here the drain is loaded by the i.f. filter and R1 is the source resistor, by-passed for signal by C1. Gate 1 is biased by R2/R3 and gate 2 by R4/R5. R6 is a 'hold-off' resistor and C2 another signal by-pass

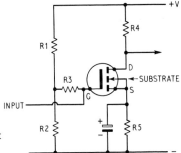

Fig. 1.25. High input impedance amplifier using insulated gate n-channel enhancement f.e.t.

capacitor. V.H.F. signal is fed in through C3, local oscillator signal through C4, while the i.f. output signal is fed out through C5. Multiplicative mixing is achieved. Fig. 1.25 shows a high input impedance (about 100 megohms) amplifier based on the insulated gate n-channel enhancement f.e.t. which, like the n-channel device in Fig. 1.24, requires a positive drain potential. The correct operating point is achieved by the potential divider R1/R2 in the gate circuit, which provides forward-biasing. Hold-off from the divider is obtained by R3 so that the inherently high input impedance is not shunted. R4 is the drain load across which the amplified signal is developed, while R5 is the source resistor for d.c. stabilisation, bypassed for signal by the electrolytic capacitor.

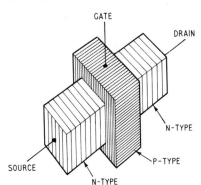

Fig. 1.26. Elementary impression of n-channel junction gate f.e.t.

An elementary impression of the n-channel junction gate f.e.t. is given in Fig. 1.26. Basically, a 'block' of n-type semiconductor passes through a section of p-type semiconductor. The p-type constitutes the 'conduction channel', thus making the device n-channel. The source is at one end of the n-type and the drain at the opposite end. The gate is formed partly by the p-type semiconductor and the resulting p-n junction is in reverse conduction because the gate (p-type) is negative with respect to the source (n-type). Hence the gate input is at high impedance.

This reverse biasing tends to inhibit the flow of electrons (negative carriers) in the n-type semiconductor. Under zero bias conditions the n-channel conductivity is high, but as the negative gate bias is increased so the number of carriers available for source-to-drain conduction diminishes, and when the bias is raised conduction eventually ceases, the channel being 'pinched-off'.

A p-channel device works similarly, but with this type the channel is fabricated from p-type semiconductor and so the electrode potentials operate in the reverse sense.

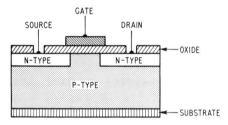

Fig. 1.27. Elementary impression of n-channel enhancement insulated gate f.e.t.

Fig. 1.27 gives an elementary impression of an n-channel enhancement insulated gate f.e.t., which has an even higher gate impedance than the junction gate type because the gate is insulated from the other elements by a thin layer of oxide. The nature of the device in this respect is sometimes exposed in the name, such as the m.o.s.t., which is the abbreviation of metal oxide semiconductor transistor.

In Fig. 1.27 the conduction channel is developed between two regions of n-type semiconductor, which are particularly rich in electron carriers, at the ends of a piece of p-type semiconductor. The gate is placed across the two n-type regions and insulated from them by the oxide. One n-type region constitutes the source and the other the drain.

Biasing in one mode is such that both the drain and gate are made positive with respect to the source (see Fig. 1.25). Electrons in the n-type source are attracted by the positively biased gate owing to the formation of an electrostatic charge resulting from the gate potential on one side of the oxide and the opposite potential on the drain and substrate. Current carriers (electrons in this case) are thus drawn into the space between the source and drain, being attracted by the latter

Fig. 1.28. I.C. used in the i.f. channel of a hi-fi receiver

since this is positively charged. The quantity of electrons in the channel, therefore, is governed by the number attracted from the source by virtue of the gate biasing. The greater the bias potential, the greater the number of electrons available for conduction.

The term *enhancement* is applicable to this type of f.e.t. because the conductivity over the source-to-drain channel is enhanced by the gate bias.

The *depletion* f.e.t. operates by virtue of an impurity concentration, such that the gate bias *reduces* the number of carriers available for conduction in one polarity, while increasing the number in the opposite polarity. Thus the gate control of the first results from current carrier enhancement and that of the second from current carrier depletion.

F.E.T.s are used extensively as amplifiers in domestic-based electronics; also as voltage-controlled resistors for various control applications. They are also sometimes employed as switches and because of their high input resistance have certain advantages over bipolar transistors. Designers are finding them desirable for r.f. amplifiers and mixers in radio receivers, particularly those for v.h.f./f.m. reception (see Fig. 1.24). Owing to their almost square law, they yield very little third-order harmonic, and this tends to minimise inter- and cross-modulation troubles, particularly in strong signal areas where bipolar transistors can sometimes result in overloading and spurious signals. More information on the receiver circuits adopting f.e.t.s is given in my *Radio and Audio Servicing Handbook*, 2nd Edition (Newnes Technical Books).

## Unijunction transistors

This is a special type of transistor which is equipped with two bases and an emitter. There is no flow of current from one base to emitter until the emitter rises to a specified potential, called the intrinsic 'stand-off' ratio, corresponding to a fraction of the voltage between the bases – commonly around 0.5 to 0.7.

As soon as the emitter voltage exceeds the 'firing' voltage the emitter current rapidly increases, then being limited by the intrinsic resistance of the circuit. Varying the voltage between the bases alters the 'firing' point.

The device is often used to trigger thyristor circuits.

## Thyristors

These devices have four semiconductor layers yielding, for instance, p-n-p-n junctions – one more junction than the transistor. However, only three terminations, labelled anode, cathode and gate, are used as shown in Fig. 1.29. For the anode and cathode to behave as an ordinary diode and pass current in the forward direction, a current must be caused to flow in the gate/cathode circuit.

After a triggering voltage applied to the gate (often from a unijunction transistor or similar device) satisfies this requirement the device continues to pass forward current until this current falls below the so-called minimum 'holding current' or, of course, the anode becomes negative with respect to cathode.

The thyristor is commonly exploited in control circuits of various types, including power supply control (in TV sets, for example), motor speed control, lamp brightness control (the dinky little knobs that we are now using instead of light switches to enhance the fire-side intimacy!) and so forth.

The thyristor is found in two types of circuit – phase control and cycle period control. With the first the thyristor is triggered on at any required point on the mains cycle or, at least, half-cycle owing to rectification so that between the triggering point and the conclusion of the half-cycle the required amount of power can be abstracted by the load. With the second, the thyristor is triggered on for a controlled number of rectified half-cycles each second, the load then responding to their 'average'. This type has longer periods between

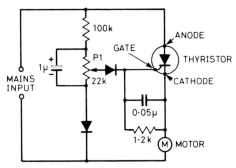

Fig. 1.29. Simple thyristor motor controller

on/off states and is less suitable than the first for quick-responding control, but it causes less mains injected interference than the phase control type.

An example of a simple motor control circuit using a thyristor is given in Fig. 1.29. As P1 is turned so the relative phase of the supply voltage, divided down at the gate, changes thereby triggering the thyristor over the required half-cycle period. It is, of course, important to ensure that the thyristor chosen for this job has suitable ratings – e.g., 340V peak or 680V peak-to-peak for 240V mains supplies! Device dissipation must also be taken into account.

### Triacs

These are essentially 'two-way' thyristors, designed for control in each direction, thereby representing a pair of thyristors.

### Diacs

These are basically bilateral switches. There is no gate electrode, but the device is triggered by the voltage across the diode rising, in either direction, to the so-called triggering voltage. The voltage across the diode falls substantially when triggered. Triacs and diacs are often found in the gate circuits of thyristors; their purpose is to achieve clean, sharp triggering pulses.

### Integrated circuits

This chapter would be incomplete without mention of the integrated circuit (i.c.). There are a great many varieties, some designed

38  Transistor fundamentals

specifically for logic and computer applications and developments suitable for use in radio and television receivers (particularly colour), stereo decoders and amplifiers. In the domestic sphere the devices are often called linear i.c.s or operational amplifers, which distinguishes them from the switching devices used in computer and other applications.

An i.c. is a complete circuit assembly built upon a chip of crystal or other substrate in virtually microscopic dimensions, and then encapsulated in one of various forms. A common form of i.c. is based on a single chip of silicon, and is then generally known as a monolithic device. This technique facilitates the development of multiple transistor (bipolar and f.e.t.) and diode (some zeners) elements on the single monolithic chip with conductor and resistor interconnections.

A single i.c., therefore, can replace a whole 'chunk' of conventional circuitry in a volume sometimes little larger than a conventional transistor. Indeed, some i.c. encapsulations are similar to those used for transistors. Some idea of this can be gleaned from Fig. 1.28, which shows an i.c. in the i.f. channel of a hi-fi receiver. Another i.c. mounting is the so-called 'flat pack', shown in Fig. 1.30 by Mullard, where the successive stages in the encapsulation correspond from left to right, starting with the top row, to the Kovar grid, base, frame and glass pre-form, gold-plated assembly, chip soldered assembly, gold wires bonded to chip, top of frame and lid, sealed flat pack and finally flat pack lacquered and complete.

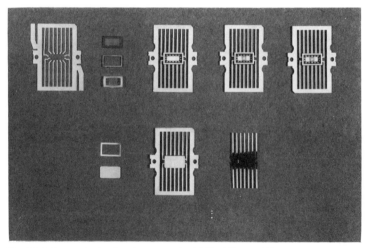

Fig. 1.30. Stages in the development of a Mullard 'flat pack' i.c.

For an i.c. to work in a complete design it needs to be connected to several external components and, of course, to a suitable power supply. The latest i.c.s used in radio receivers embody hosts of transistor and other semiconductor elements with microscopic interconnections. Some devices are even more elaborate, particularly those used in the colour sections of television receivers. Apart from the chrominance sections, i.c.s are also being used in the video, sound and intercarrier sections. These replace a sizeable part of the discrete components which would otherwise be required, and thus tend to cut the cost as well as the size of the receivers. One example of an i.c. in the sound channel is given in Fig. 1.31. The intercarrier signal is applied via a transformer at the input, and the i.c. provides the necessary intercarrier signal amplification, limiting, f.m. detection and audio amplification for driving an output transistor.

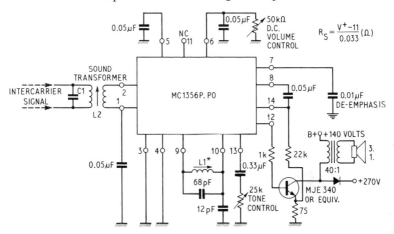

Fig. 1.31. Linear i.c. in sound section of television receiver

For basic audio applications, the i.c. might be a relatively simple device composed of a single f.e.t. and bipolar transistor with a resistor, or it might be more involved with ten (or more) transistor elements and various diodes for stabilisation and reverse-biased for capacitance, etc., such as the RCA CA3012, which is sometimes found in the i.f. channel of an f.m. tuner or the 7300 used in cassette decks for Dolby noise reduction.

I.C. design commences with full-scale layer-type drawings, and the chip is ultimately processed in a similar way to a transistor, using silicon as the crystal substrate. Fig. 1.32 attempts to expose the cross-section of a part of an i.c. The chip is around $1.5 \times 1.5 \times 0.3$

mm, which is extremely small! Photography creates a multiplicity of the same design and this is followed through to the crystal, which gives the individual chips later after separation.

In Fig. 1.32 the substrate is p-type silicon in which is buried a layer of n+-type semiconductor. The collector of the transistor element in the diagram is fabricated from an n-type epitaxial layer, and like epitaxial transistors, this layer is grown in a special reactor. Design is based, in this particular example, on the planar transistor technique, which diffuses the base and emitter planes into the epitaxial layer, as shown.

Because the epitaxial collector region is lightly doped its resistivity is high, but is reduced in practice by a small layer thickness, the stress then being borne by the substrate. The two conflicting factors for high performance – high collector resistivity to retain high breakdown potential and a low collector resistance for a low saturation voltage – are thus attained by the design.

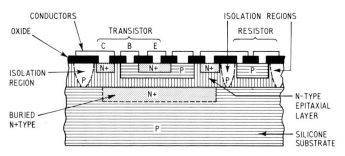

Fig. 1.32. Impression of the components of an i.c. in cross-section. See text for full description

The upper epitaxial surface is oxidised and 'windows' are produced by photolithography so that p-type semiconductor serves as 'isolating' regions between the components shown in the diagram. Isolation results from the p-n junctions between the p- and n-type semiconductors, since these are reverse-biased and therefore exhibit a high resistance.

Windows are also cut in the oxide layer for the base and emitter regions of the transistor element, developed by diffusion of the n+ collector contact region and the p-type 'resistor' area. Re-oxidising seals the windows after each diffusion. More windows are cut in the final oxide layer and a thin metallic film is evaporated over the whole surface which, when the unwanted areas are etched away, forms the conductor electrodes. The buried n+ semiconductor reduces the lateral resistivity of the collector, which helps to maintain the low saturation voltage.

Diodes are developed either as a transistor with the collector and emitter shorted or by processing as for a transistor but without diffusing the emitter to the base region. Small capacitance values are provided by reverse-biased p-n junctions, while resistors are sometimes evolved by the diffusion of p-type semiconductor simultaneously with the bases of the partnering transistor elements. Since the depth of the diffusion in this respect is dictated by the required thickness of the base region, the required resistor value is obtained by controlling the *surface area* of the p-type semiconductor, neatly achieved by suitably dimensioning the diffusion window.

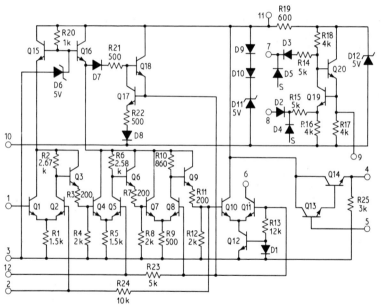

Fig. 1.33. Circuit of RCA CA3043 i.c.

Fig. 1.33 shows the circuit of the RCA CA3043 monolithic i.c., which is suitable for the i.f./dectector/a.f. stages of an f.m. receiver or tuner. The device embodies a multiplicity of long-tail pairs and diodes and separate transistor elements. These provide stages of i.f. amplification (the four-stage emitter-follower-coupled sections capable of 80 dB gain at 10.7 MHz) and limiting, f.m. detection and zener diode supply regulation, together with a.f. amplification. The f.m. section is distinguished by the circuitry which provides forward bias to the detector diodes D2/D3, and there is provision here for a.f.c. potential. The a.f. section is arranged to yield a low impedance drive for subsequent amplification, while the zener diode section gives regulated and decoupled supplies for the other sections.

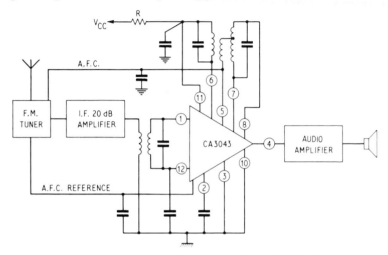

Fig. 1.34. Block diagram of complete f.m. receiver using the CA3043 i.c.

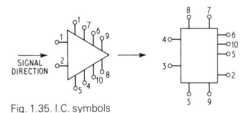

Fig. 1.35. I.C. symbols

This device in a circuit with discrete components forming a complete f.m. receiver is shown in Fig. 1.34. Typical i.c. symbols are shown in Fig. 1.3.

## More i.c.s

Pretty well every item of latter-day domestic electronics incorporates i.c.s of both the 'linear' (usually termed operational amplifier – op-amp for short) and digital (see later) kind. As has already been noted these represent 'chunks' of circuitry – some relatively simple

Semiconductors 43

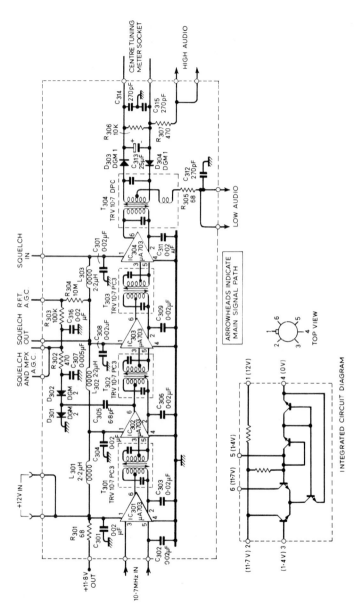

Fig. 1.36. Circuit of i.c. i.f. channel of the Scott 342-B tuner-amplifier. The circuit of each of the four i.c.s. is given in the inset below. Separate diodes for the ratio detector are used in this circuit, across the balanced load of which is connected a centre-tune meter. The arrowheads are the i.c. symbols

44  Transistor fundamentals

and others astonishingly complicated. They sometimes include linear and digital sections (e.g., in colour TV sets).

The basic op-amp has two inputs, one output and power supply inputs, with 'earth'. With some configurations a split power supply rail is desirable. This means that relative to 'earth' there are both positive and negative supply inputs. On the other hand, i.c.s will be found powered between chassis and the single supply rail. Typical in this respect are the i.c.s shown in the i.f. channel of a hi-fi receiver in Fig. 1.36, where the circuit of the common i.c. is shown separately. Here you will see the two inputs, the single output, the positive supply input and the 'earthy' terminations.

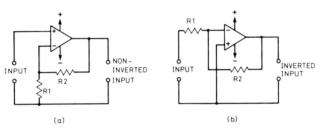

Fig. 1.37. Basic op-amp configurations. (a) non-inverting and (b) inverting. Closed-loop gain approximates R2/R1, but see text

Most op-amps are very high gain circuit 'blocks', the required operating gain being established by a feedback loop, as shown at (a) and (b) in Fig. 1.37. The inputs are labelled plus and minus because the first does not invert the signal while the second does (non-inverting and inverting inputs). Both inputs provide gain, but it is the difference signal between the two inputs which appears in amplified form at the output.

Parameters of op-amps include open loop voltage gain (the intrinsic gain of the device which is usually very high allowing feedback to tailor the requirement), nominal supply voltage, output and input impedances, input offset voltage and current, output voltage swing at the nominal supply voltage, slew rate (the maximum rate at which the output voltage can slew or change), open-loop bandwidth (a function of slew rate) which is widened with feedback and operating temperature range. Some i.c.s need to be mounted on heat sinks.

The offset voltage (or current) refers to the difference in voltage (or current) between the two inputs to provide zero d.c. at the output. The external circuit may include a small preset for setting the output d.c. to zero.

Output voltage is commonly given as ±peak voltage referred to the output load resistor.

Another parameter is the common mode rejection ratio. This refers to the effectiveness of the input balancing and expresses as a dB ratio the rejection at the output of two signals applied together at the two inputs. This is an important requirement for reducing the sensitivity of the device to extraneous mains fields, for example.

Basically, assuming that the open-loop gain of the op-amp is very high and that R2 is not any more than about 100 times greater than R1, then the closed-loop gain (with feedback applied) approximates the ratio R2/R1 (Fig. 1.37).

## Digital i.c.s

A digital i.c. is essentially a switching device where the switching and counting input consists of 'pulses'. The output thus goes either 'high' or 'low' depending on whether the i.c. is switched off or on. The on/off levels are coded 1 and 0 (it really doesn't matter which way round they are) and the counting and controlling are related to figures of the binary scale, which involve merely the digits 1 and 0. A greater advantage of this method of operation is that signal noise is no longer troublesome since the tops and bottoms of the pulses can be chopped off to eliminate any noise if it is present.

It is not possible to delve into the binary code in the compass of this book, but to give some idea of the principle, decimal 1 corresponds to binary 1, decimal 2 to binary 10, decimal 3 to binary 11, decimal 4 to binary 100, decimal 5 to binary 101, decimal 6 to binary 110 and so forth.

Apart from control operations, ordinary analogue signals (such as those from audio discs, say) can be converted to digital form by an analogue-to-digital converter, processed or even transmitted without the normal noise problems and then changed back to the analogue form by a digital-to-analogue converter.

There are already hi-fi amplifiers which adopt this principle, and the transmitting authorities are using digital circuitry on link circuits. The process is also used for digital audio and video discs. A large slice of the future of radio, television, audio and video will be very closely geared to digital systems.

At the present digital i.c.s are mostly used in domestic electronics for control functions. For example, the remote control arrangements of TV receivers, hi-fi amplifiers and receivers and audio and video cassette decks are all based on digital electronics. The circuit path from the hand-held remote control unit is often established by an infra-red link. The press of a button on the control unit applies the appropriate digital encoding to a special infra-red-emitting LED. The signal is picked up by an infra-red-responding photo-diode at the receiver which, after being processed by the digital i.c.s, then activates the desired function.

Digital circuits are also found in some of the more sophisticated hi-fi receivers and tuners for tuning control and frequency readout (a digital display being used for the latter). In these cases the local oscillator signal may be digitally synthesised against a reference crystal so that the tuning operates in small synthesised steps. In less elaborate designs the digital electronics may be mostly concerned with the digital readout of frequency.

Station recall becomes readily possible by the use of digital 'memory blocks'. The tuned frequency proper is handled by capacitor diodes (page 27), the digital electronics merely setting the voltage required by the capacitor diodes to tune to a particular frequency.

## Servicing hints

It is obviously not possible to change faulty parts in an i.c., so if an i.c. is suspected of being responsible for a fault condition it is often easiest in the long run to try a replacement! However, before this is done it certainly pays to make a number of voltage tests around the i.c. Typical troubles are resistors which have changed value, leaky capacitors and collapse of supply voltage.

A quick-cut operation resolves to isolating the suspect stage to see whether a signal can be passed round it, thereby proving the suspect and that the other parts of the circuit are operating. As an example, let us suppose that the i.f. channel in Fig. 1.36 is defunct, this being proved by lack of audio output signal when a suitably modulated 10.7 MHz signal is applied at the input. A swift way I have found of locating the stage responsible is by bypassing each i.c. in turn with a 100 pF capacitor. When the defunct stage is so bypassed signal continuity will be restored but the gain of the channel will be considerably less than it would be with the stage active. This should not bother us because all we want to measure or hear is just a trace of signal.

In the case of the i.c.s in Fig. 1.36 the bypassing should be between pins 3 and 6 (input and output). The same artifice can be employed to bypass suspect stages of a transistor circuit; but the value of the bypass capacitor chosen should relate to some degree, anyway, to the circuit impedances and frequency of the signal. With r.f. stages the tuned filters or transformers would be subjected to detuning by the presence of the bypass capacitor; but this is not important provided some signal gets through! An advantage of the scheme is that it can be practised without any test instruments whatever, provided there is an input signal (from an aerial or audio source) and a speaker to hear it by. Just couple in the capacitor using crocodile clips, turn up the volume and listen for signal (in the case of a receiver, of course, you will need to tune a station in the normal way).

When you get signal continuity, the stage bypassed should be scrutinised. if you find no problems with the d.c. conditions then about the only thing left to do is to replace the i.c.

Some i.c.s are plugged into holders, which certainly aids servicing. Sadly, many are soldered directly into the p.c.b. and are much more difficult to extract. Frankly, the best way of extraction is to snip the wires close to the p.c.b. and wriggle the wires out of the board with a soldering iron and desoldering tool. Always use the least amount of heat and blow on the p.c.b. after each operation to help cool it. Remember that the terminations are connected to the silicon chip through fine gold wires and that destruction can result if the heat is left on too long. Indeed, quite a few i.c. troubles can be traced to poor welding of the wires inside; but there is nothing you can do

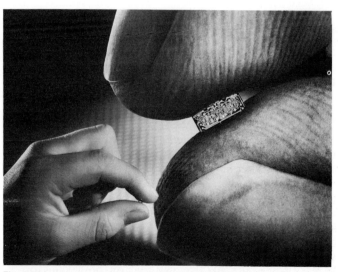

Fig. 1.38. Just how small integrated circuits really are is shown by this picture of an unencapsulated Mullard TTL decade counter containing over 120 components. Decade counters are used in computers, desk calculators, adding machines and all types of process control equipment (Mullard photo)

48    Transistor fundamentals

about this! Such faults, though, usually manifest within a few months of equipment use.

If you are not sure that the i.c. is defective, then you will have to extract it without wire cutting. The latest desoldering tools or braid tend to ease this exercise.

After a faulty i.c. has been located always check the surrounding circuit to make sure that trouble here was not responsible for its demise before switching on after inserting a replacement. It goes without saying that all power should be removed from the equipment before changing an i.c. or any other component for that matter!

Some i.c.s are sensitive to static, especially the MOSFET input ones (and this also applies to MOSFETs), so the shorting links fitted should not be removed until after the device is well and truly in circuit.

It is sometimes possible to use a quasi-equivalent i.c. if the actual device is no longer available. While this may work it might give rise to instability or some other odd effect due to a wider bandwidth, for example. This can generally be resolved by a little capacitive loading at the output (a pF or two may be enough).

The photograph in Fig. 1.38 shows just how small an i.c. chip might be.

## Microcircuits and hybrids

Distinct from the monolithic i.c., there are other types of microcircuits, called thick films and thin films. These are processed on high-grade grade glass or glass-coated ceramic, with the circuit in the case of the thick films being 'printed' on the substrate, which takes the form of a very thin wafer, and then fired. The technique involved is after the style of printed circuit design and manufacture, but on a much smaller scale. A number of circuits are generally processed in one go and then later, divided for individual application.

Thin films, on the other hand, are evolved by evaporating them on to the substrate in a vacuum. Resistor elements can be made by the evaporation of nickel-chromium alloy, while the actual conductors are made either by the evaporation of gold or by the depositing of nickel on to the films, depending on the requirement. Resistors are formed by zig-zag patterns, which are sometimes computer tailored to provide a very high degree of accuracy, while capacitor elements are made by evaporating suitable material on both sides of the substrate, the substrate in this case then acting as the dielectric.

There are hosts of specialised procedures adopted, and quite complex circuits can be produced, including components. The hybrid

Semiconductors 49

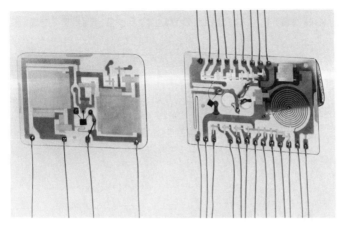

Fig. 1.39. Examples of Mullard linear thin-film circuits

microcircuit includes active devices, such as transistors and even monolithic i.c.s. The number of discrete components required for a given application can be reduced by the hybrid technique, and devices capable of handling greater power than monolithic counterparts become possible. Already hybrid circuits are appearing in domestic-based electronics, such as hi-fi amplifiers, and their numbers are almost certain to increase rapidly as times goes on.

Fig. 1.39 gives two examples of Mullard linear thin-film circuits. That on the left of the picture is a 50 MHz oscillator, the crystal being mounted on the back of the substrate, while that on the right is a single-stage wideband amplifier responsive over 0.5–70 MHz.

# 2 Preliminary circuit and transistor tests

Aside from circuit and mechanical troubles, transistor equipment is prone to two major fault possibilities. These are (i) transistor failure and (ii) power supply failure. Generally speaking, resistors, capacitors, inductors, transformers and so forth are subjected to a smaller operational stress than their counterparts in valve equipment where the power and the temperature are relatively much higher.

Transistor components usually have voltage and power ratings below that of those employed in valve circuits, but in spite of this their life expectancy is above that of valve-type components. However, they are somewhat more susceptible to both electrical and mechanical damage.

The nominal power consumption of transistor equipment is usually small when compared with that of battery-operated valve equipment. This means that the transistor battery is expected to have a relatively long life. The battery may thus be overlooked when the equipment ceases to work and tests of a deeper significance may be contemplated. However, in many cases, waning battery power is exhibited by characteristic symptoms which we shall investigate later.

**Transistor reliability**

It is perfectly true that transistors on the whole are more reliable than valves. That is, once they have settled down in the equipment. They may then continue to work at optimum efficiency for many years. As is well known, valves diminish in efficiency due to emission deterioration, which is a thermal problem and does not apply to the transistor, which has no heater.

Preliminary circuit and transistor tests 51

The worst time for transistors is during their very early life in the operating equipment, and it is during this period that the majority of failures are observed. After this period, the failure rate falls to a very low and constant level until the devices eventually suffer from old age, when the failure rate commences to rise again.

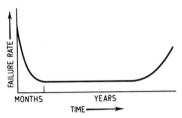

Fig. 2.1. The life curve of a transistor. After an intially high failure rate, the rate diminishes rapidly and settles down to a consistently low level until the transistor fails due to very old age

It is virtually impossible to put an accurate time-scale to the transistor life curve owing to the lack of statistics. However, the general idea is shown in Fig. 2.1, where the time-scale is calibrated roughly in months and years.

## Power supply test

From the foregoing, therefore, it is revealed that defunct transistor equipment should be subjected to two immediate tests – one of the power supply and the other of the transistors. A multirange testmeter is all that is required for the power supply test. First the voltage should be checked with the equipment switched on (that is, with the battery under load) and second the total current consumption should be checked by introducing a suitable current meter is series with the power supply. This is accomplished by disconnecting, say, the

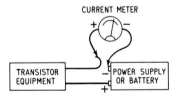

Fig. 2.2. The quickest way of checking total current

negative battery lead and putting the negative of the meter to battery negative and the positive of the meter to the equipment negative lead. The idea is shown in Fig. 2.2.

The instrument should be switched on to a higher than expected current range before switching the equipment on, and then when a reading has been obtained on the higher range the instrument can be switched down as required. This will prevent meter damage due to a short in the equipment. It is also best always to disconnect the power or switch off the equipment before adjusting the range of a testmeter.

There is often a high 'kick' of current when the equipment is first switched on owing to charging electrolytic capacitors.

Although the magnitude of the total current consumption of the equipment may not appear to be all that important at this stage, the fact that the equipment is taking power means, at least, that the power supply feed circuits are correct, including the on/off switch and fuse if fitted.

One will almost certainly have a rough idea of the range of current that the equipment should be taking. If the current is zero, one would conclude that something is amiss with the power feed circuits (i.e., switch, fuse or wiring faulty) since most transistor equipment, even if only a single transistor is employed, has a potential divider across the battery supply which would pass a very small current in any event.

An abnormally high current reading would indicate a short or a leak in the equipment. The desirability of the initial power supply test is thus revealed, for it allows us to find out quite a bit of useful information about what is happening in the equipment before it is taken from its case or cabinet. We may find that to bring the equipment back into service we need only to replace the battery or fuse or resolder the battery connector or lead.

Assuming that the power supply test tells us that the fault is caused by trouble actually inside the equipment, then, to follow the logic formerly outlined, we need next to discover how the transistors are behaving d.c.-wise in the circuit.

## Operating conditions

From Chapter 1 we discovered that in almost all transistor circuits the base needs to be biased for forward current and the collector in the reverse current direction. The base bias in practical circuits is established either by connecting a high value resistor between the base and the power supply line (negative for a p-n-p transistor and positive for an n-p-n type) or by the use of a resistive potential divider across the power supply with the junction connected to the base. These two arrangements are shown respectively at (a) and (b) in Fig. 2.3. Here the transistor is the p-n-p type, but exactly the same principles apply with n-p-n transistors except that the supply polarities are reversed.

In an amplifier, the base bias sets the standing collector current for the required operating conditions. For instance, for class A the bias is set for an $I_c$ which will swing up and down equally due to the alternate half cycles of input signal current injected into the emitter junction. If the base bias is too high the transistor may *bottom* (i.e. saturate, see page 75) when the input signal swings negative at the base while if it is too low collector current cut-off may occur when the signal at the base swings positive (i.e., on the positive half cycles). The correct base bias is thus important.

For class B operation, the base bias is set more towards collector current cut-off. In such a circuit two transistors are employed and the arrangement is for one transistor to rise into collector current while the other is switched off on one half-cycle and for the action of the transistors to change over on alternate half-cycles.

For class C operation, the transistors are biased for zero $I_c$, collector current flowing only during signal swings. The condition between class A and class B is called class AB which, again, of course, is established by the base bias.

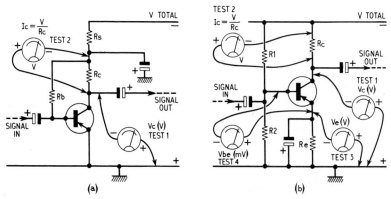

Fig. 2.3. These circuits show the points at which useful test readings can be taken: (a) in simple stage with no base potential divider and (b) in the common stage where a base potential divider is used

We shall be considering the signal conditions of transistors in detail later, and for the time being all we need to understand is that for most basic amplifier applications the base is biased for a standing $I_c$.

Now to revert to Fig. 2.3. The base bias in circuit (a) is established by $R_b$. The $I_b$ is essentially equal to the voltage at the junction of $R_c$ $R_c$ divided by $R_b$. The $I_b$ in µA is equal to the voltage divided by $R_b$ in megohms. Thus, if $V$ is 10 and $R_b$ 1M, $I_b$ will be about 10 µA.

The circuits in Fig. 2.3 are in the common-emitter mode. Thus, by looking at the common-emitter transfer characteristics for the type of transistor used (see, for instance, (a) in Fig. 1.18) we can glean some idea of the $I_c$ that any value of $I_b$ is likely to raise. Since we are concerned with fault diagnosing as distinct from design, we need not bother too much about precise current values, and we can be assured by the knowledge that the circuit was working correctly before it went wrong! Our job is simply to find out what has gone wrong.

$R_c$ in Fig. 2.3 is the collector load resistor while $R_s$ in the circuit at (a) is for d.c. stabilisation, noting here that there is no emitter resistor. Let us suppose that the overall $I_c$ is tending to rise due to temperature increase. The current in $R_s$ rises likewise. The voltage dropped by this resistor also rises and this results in a fall in voltage at the junction of $R_s R_c$. This means that the negative base bias is pulled back, the action of which causes a drop in $I_c$ resulting from the transistor effect. The circuit is, therefore, stabilised.

Now let us look at the circuit in Fig. 2.3(b). Here the base bias is set by the negative voltage tapped down the potential divider R1 R2 and by the voltage dropped across the emitter resistor $R_e$. The forward current in the emitter/base junction is thus determined by the voltage between the emitter and base ($V_{be}$). We will recall from the last chapter that the input characteristic (Fig. 1.16) shows how the $I_b$ rises with increase in $V_{be}$ (note that $V_{be}$ is sometimes expressed as $V_b$). As an illustration, the small signal AF114 Mullard transistor produces an $I_b$ in the region of $24\,\mu A$ for a $V_{be}$ of $-300\,mV$. Across the emitter/base junction of the same device would be developed approximately $270\,mV$ for an $I_b$ of $10\,\mu A$.

In the circuit under discussion $R_e$, is the 'stabilising' emitter resistor. As mentioned in the previous chapter, the action is such that any increase in $I_c$ due to temperature rise flows also through $R_e$, giving a corresponding increase in voltage drop. This causes the emitter to rise more negatively with respect to the base, which is the same as the base going less negative, causing $I_c$ to be pulled back.

$R_c$ is the collector load resistor. The collector load, of course, could be the primary of a transformer or an inductor, but a pure resistive load is given in the circuits of Fig. 2.3 as these are representative of simple amplifiers. To avoid degenerative current feedback, $R_e$ is capacitively bypassed, the reactance of the capacitor being low at the operating range of frequencies in comparison with the value of $R_e$. In some circuits negative feedback is deliberately applied either by introducing a small unbypassed $R$ in the emitter circuit or by leaving the capacitor out altogether. Negative feedback tends to increase the input impedance of the stage, while also reducing its gain.

Degenerative feedback is avoided in the circuit at Fig. 2.3(a) by the connection of a bypass capacitor between the junction of $R_s R_c$ and the supply positive line. $R_s$ and the electolytic capacitor can also be considered as 'decoupling' components.

At this juncture, it should be noted that either the negative or positive power supply line can be made 'earthy'. In the circuits shown, however, the positive line is the earthy one. We shall see later that the configuration common-emitter, common-base or common-collector) is determined signal-wise by 'earthing' the common electrode. Since the circuits in Fig. 2.3 are of the common-emitter mode, the emitter is earthed at (a) direct and (b) signal-wise through the emitter bypass capacitor. Input is applied at the base and taken from the collector.

## Capacitor values and polarisation

The relatively low or medium input and output impedances involved demand coupling capacitors whose reactance at the operating frequencies is low in comparison. Should the reactance of a coupler be greater or not all that much lower than the transistor input or output impedance excessive attenuation of the lower frequencies will result. It is for this reason that high value electrolytic capacitors are employed in transistor a.f. amplifiers. In h.f. amplifiers the value of coupling capacitors is lower, since the capacitive reactance is lower at signals of those frequencies.

While we are on this subject the question of the polarisation of the electrolytics should be considered. The capacitors are connected in circuit so that they are always correctly polarised. That is, the positive side should be connected to the most positive side of the circuit. Electrolytics used for coupling one stage to another should be connected to n-p-n transistors so that the positive side is on the collector of the preceding transistor and the positive side on the base of the following transistor. The polarity, of course, is opposite on p-n-p transistors.

The input capacitors in Fig. 2.3 circuits are not connected in this way for here it is assumed that a preceding stage is not employed, in which case a load with d.c. continuity may be connected between the positive plates and the positive supply line. Polarisation of bypass and decoupling capacitors is more obvious.

It is extremely important that replacement electrolytics be connected the right way round. If inadvertently connected backwards leakage current could flow which would disturb the biasing of the transistors and make fault diagnosis more difficult.

## Circuit analysis

We now have sufficient information to establish the basic operating conditions of any transistor stage and also to acquire a fair assessment of the condition of the transistor. Although mentioned earlier that the second most important thing to do in fault diagnosing is to find out the condition of the transistors, this was not meant to imply that each transistor should be extracted from circuit and connected to a transistor tester. Indeed, extracting a transistor from some types of transistor equipment can be precarious and hence best avoided until one is almost certainly sure that the transistor is faulty. The exercise, then, lies in discovering the condition of the transistor while it is in situ.

This is achieved by making voltage and current measurements in and around the suspect stage. Assuming that the power supply is in order, the first to check is collector voltage, Test 1 in Fig. 2.3. If there is $I_c$, the resistors in the collector circuit will drop some of the supply voltage and the testmeter will show a voltage below that of the supply. If the collector load is an inductor or winding on a transformer, the d.c. resistance may be only a few ohms, in which case the voltage dropped would be low and $V_c$ would read almost the same as $V_{total}$.

If there is no voltage on the collector and the equipment is not passing an abnormally high current (determined by the first test of the power supply), then there must be an open-circuit in the collector load or in a feed resistor.

## Checking current without breaking circuit

If $V_c$ is present, the next move is to check $I_c$. The obvious way to do this is to break the circuit at the top of $R_c$ and insert a current meter. This is difficult in printed circuits and a far better way is to measure the voltage developed across a resistor in the collector circuit and from this and the knowledge of the resistance value calculate the current flowing. This is not difficult and it certainly aids towards speedier fault diagnosis. Let us take an example.

Suppose that $R_c$ is 1k (i.e., 1000 ohms) and that across it we measure 1 volt. From Ohm's law ($I$ equals $V/R$) we can discover the current. In milliamperes it is equal to the voltage measured in volts divided by the resistance in thousands of ohms (i.e., kilohms). Thus, in the example we have 1/1 or 1 mA.

The accuracy of this computation depends upon the accuracy of the resistor and the voltmeter; also upon the sensitivity of the voltmeter.

It will be understood of course, that a voltmeter of mediocre sensitivity will act as a significant shunt resistance across $R_c$, for example, and reduce the resistance value across which the voltage is developed. However, for all practical fault-finding activities the resultant current so derived is sufficiently accurate. This method of $I_c$ measurement is shown by Test 2 in Fig. 2.3.

We should now have a reasonable idea of how the transistor is behaving. If both $V_c$ and $I_c$ are normal for the type of circuit under test, we could conclude that $I_b$ is flowing and that $I_c$ is due to the normal transistor action. We may be wrong in this conclusion, however! This is because $I_c$ may be leakage current or a current due to a transistor defect. We really want to be sure about this.

We can be absolutely sure that the $I_c$ is resulting from the normal transistor effect, signifying that the transistor is, in fact, working correctly, by varying $I_b$ very slightly while we are observing $I_c$ (that is, the voltage measured across $R_c$).

$I_b$ can be varied either by altering the value of $R_b$ in circuit (a) or by altering the $V_{be}$ in circuit (b). The quickest way of handling circuit (a) is simply to shunt $R_b$ with a high value resistor. If $R_b$ is, say, 470 k, then we could shunt it by a resistor of about 1 M without pushing a destroying current into the emitter/base junction. On a 10-volt power supply this would increase the $I_b$ by about 10 µA.

## Checking beta

If the transistor is working with an $I_b$ of, say, 20 µA and an $I_c$ of 2.8 mA, then an extra 10 µA of $I_b$ should push the $I_c$ up to about 4.2 mA, depending upon beta. We can find out the actual beta from the common-emitter transfer characteristics, and then compare with the beta indication that we obtain from the tests. Beta is equal to the change in $I_c$ divided by the change in $I_b$. In the above example, it would be 2 000/10 or 200.

We must take care to avoid over-running the collector when making a test of this kind. Too great an $I_b$ increase might ruin the emitter/base junction or destroy the collector junction by causing the collector dissipation to be exceeded. With the Mullard AF114 the dissipation limit at 45° C is 50 mW. This corresponds to an $I_c$ of 5 mA when the $V_c$ is 10V.

Beta in a circuit of the kind shown in Fig. 2.3(b) is more difficult. However, it is possible to tell whether the transistor is working by shunting R2 with a resistor of approximately the same value as R2 itself. This will *reduce* $V_{be}$ and thus a drop in $I_c$ will be observed in a circuit in which the transistor is in working order.

58  Preliminary circuit and transistor tests

In circuits where the $I_c$ fails to change due to a change in $I_b$ or $V_{be}$, something is almost certainly wrong with the transistor. However, before attempting to extract the transistor for isolated test or try a replacement in the circuit, one or two confirming tests should be undertaken.

In a circuit such as Fig. 2.3(b), where there is an emitter resistor, $V_e$ can be measured, as shown by Test 3. $I_e$ can be assessed in a similar way to $I_c$ by dividing the voltage measured by the value of the emitter resistor in kilohms to get the answer in milliamperes. Thus, 1 volt across 1 k would signify an $I_e$ of 1 mA.

### Shorting collector junction

If there is $I_c$ but no $I_e$ (that is, zero $I_e$) there would almost certainly be a short or leak in the transistor collector junction. Under this condition $I_c$ would not change with change in $I_b$ or $V_{be}$.

In Chapter 1 we learnt that the $I_e$ is the $I_b$ above the $I_c$. This is because both the $I_c$ and the $I_b$ flow in the emitter circuit. Careful readings have to be taken to observe this difference between $I_c$ and $I_e$, since $I_b$ may only be a few μA. Should the $I_c$ and the $I_e$ readings seem pretty normal and yet a change in $I_b$ or $V_{be}$ fails to cause a change in $I_c$, it may be as well to make a check of $V_{be}$ right at the transistor wires, as shown by Test 4 in the circuit at (b). A very sensitive meter is needed, of course, to indicate a few millivolts.

If the reading is normal and it changes when R2 is shunted, for instance, and yet $I_c$ still holds steady, the transistor is definitely faulty.

Should there be no $I_c$ or $I_e$, the $V_{be}$ test should be made again. If there is no $V_{be}$ indication, the meter test prods should be moved, one at a time, to the actual circuit from the transistor wires. The reason for doing this is that sometimes a dry joint or poor soldered connection develops between the transistor wires and the circuit conductors, especially on printed circuit boards. The connections may appear to be well made with nice blobs of solder, yet beneath the blobs may exist very poor electrical connections. Such trouble would starve the transistor of base current and thus kill the collector current.

### Open-circuit collection junction

If lack of $I_c$ is caused by an open-circuit collector junction this would be revealed by the $V_e$ test (Test 3) showing only a very small voltage, owing to the *base current* alone causing a volts drop across the emitter resistor.

By the four tests described, much about d.c. operating conditions of the transistor in circuit can be discovered. At the least, we would then have a case to extract the transistor from the circuit for a more detailed test or for trying a replacement.

One or two words of warning about the vulnerability of transistors, particularly the small ones, would not be amiss. Transistors do not like peaky voltages or currents, called voltage or current transients. Although these are very short lived they can destroy a base or collector junction completely or severely alter its characteristics.

Transients can be produced by connecting mains-operated test equipment to the various circuits without adequate isolation. Indeed, it has been known for the normal extremely small leakage between the element and the case of a soldering iron to produce a current transient on contact with the circuit of sufficient magnitude to blow the emitter junction, this usually being the weakest junction of the two. This might not have happened had the iron had been connected to a good earth, since then the leakage currents would have been bypassed away from the circuit. However, to be safe a battery powered iron might be justified.

Valve voltmeters, if incorrectly applied, can cause trouble with transistors. it is best to connect such mains-operated instruments to the circuit before switching the instrument and the transistor equipment on. Signal generators, and like equipment, if mains-operated, should be connected through isolating capacitors – one in each lead – of the lowest possible value consistent with the application in hand.

Dangerous transients can also be created by connecting and disconnecting transistors and components when the circuit is under power. Always switch off first, make the connections or disconnections, and then switch on. With f.e.t.s and i.c.s containing f.e.t.s also avoid static electricity.

## Open-circuit emitter junction

An emitter junction blown due to a transient would be revealed by a complete lack of $I_e$ and a very small value of $I_c$ due to normal leakage current, i.e., $I_{cbo}$.

Transients can also cause junction shorts or partial shorts. A destroyed collector junction, for instance, may show as a higher-than-normal $I_c$ which fails to change with change in $I_b$ or $V_{be}$.

A short or partial short in the emitter junction would give an abnormally high $I_e$ and zero or very low $I_c$, the latter being $I_{cbo}$.

Under certain conditions, transistors can be ruined by incorrect testing. Most inexpensive testers check beta and collector leakage over a range of voltages and currents. Provided the instruments are operated in strict accordance with their instructions all is well.

However, by running with $I_c$ or $V_c$ too high or with too great a value of $I_b$ the total dissipation may be exceeded. This would cause the junction to overheat and increase the dissipation even further, thereby producing conditions for thermal runaway. The aim, then, should be to test the transistor with the smallest dissipation consistent with the measurement, keeping the transistor under test for the shortest possible time, and switching off quickly if the reading tends progressively to rise.

## Spurious oscillation

V.h.f. and u.h.f. transistors can themselves produce spurious oscillations in certain transistor testers. Such oscillations are revealed by a hard full scale beta deflection on the meter and by the reading falling right back when the base or collector wire on the test transistor is touched with a finger or screwdriver blade. Spurious oscillations are encouraged by the use of long connecting leads between the instrument and the transistor under test. When testing u.h.f. and v.h.f. transistors the lead-out wires should be connected direct to the terminals of the instrument. If oscillation is still troublesome it can usually be quenched by threading small ferrite beads on to the transistor wires. The beads increase the inductance of the leads which makes spurious oscillation more difficult.

Apart from disturbing the test readings, spurious oscillations can, in certain cases, destroy the transistor under test by causing an excessive rise in $I_c$ or $I_b$ (or both) and the transistor to exceed its maximum limit of dissipation.

## Multi-stage equipment

Transistor equipment usually contains more than just a single stage of the type shown in Fig. 2.3. In a simple transistor radio, for instance, there are at least six stages with transformer and capacitor coupling. These are (*i*) the frequency changer, (*ii*) i.f. 1, (*iii*) i.f. 2, (*iv*) detector, (*v*) audio amplifier or driver and (*vi*) output stage.

Although each stage can be analysed in the way that has already been explained, it sometimes happens that the stage coupling develops a fault which reflects as a fault condition in the coupled stages.

Preliminary circuit and transistor tests 61

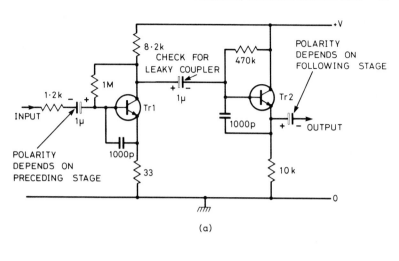

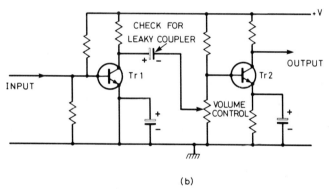

Fig. 2.4. Two stage a.f. amplifiers, (a) without volume control and (b) with volume control, both using capacitive coupling. See text referring to testing for 'leaky' couplers

Fig. 2.4(a) shows a two-stage capacitively coupled amplifier (could be a microphone amplifier). Tr1 is in common-emitter mode and Tr2 in common-collector (emitter-follower) mode. Both stages are biased for class A low-level working by the high value resistors connected between base and collector or supply line. Amplified signal at Tr1 collector is coupled to the base of Tr2 by the 1μF electrolytic.

Should the coupling capacitor develop a short or 'leak', analysis of stage Tr2 would reveal an abnormally high $I_c$. This, of course, is because the capacitor leakage current would flow into Tr2 base.

Owing to the greatly disturbed biasing conditions such a fault might well 'block' stage Tr2 completely or, if the leakage current is not severe, result in a distorted output signal.

### Check for leaky coupler

A quick way of checking for capacitor leakage is to meter the $I_c$ of the coupled stage (e.g., Tr2 in Fig. 2.4(a)), noting any change when the coupling capacitor is disconnected at one end, taking account of charge and discharge currents. If there is an appreciable change which holds steady when the coupler is reconnected, then you can be pretty sure that the capacitor is in need of replacement.

Assuming that a moving-coil microphone is connected to the input of Fig. 2.4(a) and the input coupling capacitor goes 'leaky', the base bias of Tr1 would be troubled. Since the bottom end of the microphone would be at chassis potential, the leak would shunt the base bias (assuming a moving-coil microphone which has a relatively low winding resistance) and the stage would be cut-off.

Fig. 2.4(b) shows a capacitively coupled pair of stages including a volume control. How the bias of Tr2 in this case would be affected by a leaky coupler would depend on the setting of the control. With the control fully clockwise (slider at Tr2 base end of the control) current through Tr1 collector resistor would flow into Tr2 base and significantly increase its bias, possibly causing 'bottoming'. With the control fully anticlockwise the leakage current would flow direct to chassis and not affect Tr2 biasing. Thus, the metering of $I_c$ of Tr2 would show a significant variation as the control is turned. Symptoms would be an increase in distortion as the control is advanced, almost certainly accompanied by bad crackling, and eventual cessation of signal path continuity. If the coupler leak is bad, then Tr1 $V_c$ would also fall as the control is retarded.

The record head of a tape recorder might be coupled to the recording amplifier collector through a capacitor, as shown in Fig. 2.5. A leaky capacitor here would result in a flow of d.c. through the head, which is a distinctly serious state of affairs. The pole pieces would tend to acquire residual magnetism which, if the same head is also used for replay, would impair the signal/noise ratio and upper frequency response.

The circuit also shows that the h.f. bias, required for the lowest possible recording distortion and maximum exploitation of the l.f. attributes of the tape, is fed to the head through a capacitor and bias

Preliminary circuit and transistor tests 63

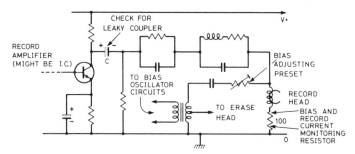

Fig. 2.5. A leak in C in this circuit would cause d.c. to flow in the record head with a consequent rise in noise and impairment of upper frequency response resulting from residual magnetism of the head pole pieces (see text)

level adjusting preset resistor from an isolated winding on a transformer. Leakage in this coupler, therefore, would not result in a flow of d.c. through the head. Recording current and h.f. bias current also flows through the 100-ohm resistor in series with the head. The h.f. voltage developed across this is often used as a measure to set the bias preset, a typical value being 30mV across the resistor for ferric HL tape. With the meter switched to d.c. there should be no trace of a reading, a contrary situation proving that d.c. is, in fact, flowing through the head.

## Direct coupling

Many circuits are coupled without capacitors, and we have already seen one example in Fig. 1.20. The transistor elements in i.c.s are invariably coupled directly. Another illustration of direct coupling is

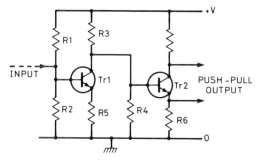

Fig. 2.6. Illustration of direct coupling, where Tr1 drives a phase splitter stage for coupling to a pair of push-pull output transistors

given in Fig. 2.6. Here the base current for Tr1 is derived from the potential divider R1/R2, while that for Tr2 is from the divider formed by R3/R4. It will be appreciated that the resistors must be chosen with care to suit both the transistor operating requirements and the

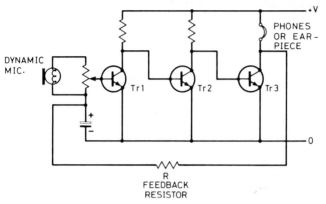

Fig. 2.7. Three-stage directly-coupled high-gain amplifier, where d.c. stabilisation is achieved by feedback from Tr3 collector to Tr1 base through R. The i.c. 'gain block' now generally takes the place of this type of circuit

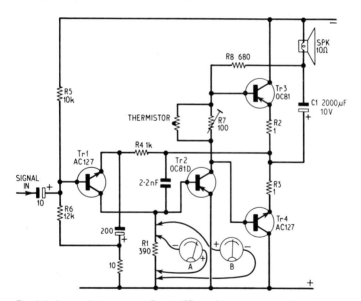

Fig. 2.8. A complementary audio amplifier using two p-n-p and two n-p-n transistors with a 'single-ended push-pull' output stage. This is fully described in the text

direct coupling. It is not possible to delete the coupling capacitor in a circuit which is not designed for direct coupling. We have already seen what a capacitor 'leak' can do!

Stabilisation in Fig. 2.6 is provided by the emitter resistors R5 and R6, and an arrangement is commonly included in the subsequent circuitry to achieve the required base current 'balance' of the driven push-pull output transistors.

Another directly coupled circuit is given in Fig. 2.7. This is designed for high gain and low noise and is the kind of amplifier which might be found in a hearing aid. Supply voltage and collector resistors are selected to yield the required overall class A amplifier condition, and in this case stabilisation is provided by d.c. feedback from the collector of Tr3 to the base of Tr1 through feedback resistor R. In general, circuits of this kind are now being replaced by the high-gain, low-noise i.c.

An illustration of full direct coupling in a practical audio amplifier is given in Fig. 2.8. Here we have a mixture of p-n-p transistor working with an n-p-n counterpart and called a complementary pair.

In addition to the complementary pair, Tr3 and Tr4, in Fig. 2.8, there are also an n-p-n preamplifier, Tr1, and a p-n-p driver stage, Tr2.

## D.C. conditions in complementary circuit

Tr1 is effectively in the common-emitter mode, with the input signal applied to the base and the output taken from the collector. Direct coupling from Tr1 collector to Tr2 base is adopted. The collector of the n-p-n Tr1 is connected to the positive source via the load R1 and the emitter is returned to the junction of R2 R3, via R4, this being the 'mid-voltage' point. A small positive voltage, relative to the emitter, is applied to the base of Tr1 from the base potential divider, R5 R6.

The collector of Tr2 is connected to a negative source, via R7, the shunt thermistor, R8 and the speaker, while the emitter is returned to the positive line.

D.c.-wise, the complementary output transistors Tr3 and Tr4 are connected in series, with the collector of the p-n-p Tr3 to the negative line and the collector of the n-p-n Tr4 to the positive line. The two emitters are then connected together through R2 and R3.

Now, the output transistors are biased towards class B; this means that the p-n-p Tr3 must have a small negative voltage at its base with respect to its emitter, while the n-p-n Tr4 must have a small positive voltage at its base with respect to its emitter. These conditions are necessary, as we have already seen, to put a little forward current in the emitter/base junctions.

These bias requirements are reflected from the directly coupled driver Tr2. In Fig. 2.9 a simplified driver/output circuit is drawn, on which voltages are given at various points. Although these are not typical, they do at least illustrate how the potentials settle down to give the required conditions. Note that the voltages are relative to battery negative.

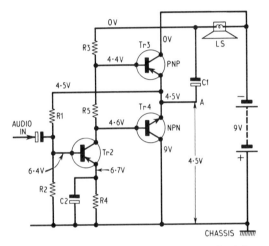

Fig. 2.9. Basic circuit of the driver and push-pull pair from the circuit in Fig. 2.8

Starting at the driver stage, we have zero volts at the top of the load R3 (actually there would be a very small voltage here due to the volts drop across the speaker winding), 4.4 V at the junction of R3 R5, 4.6 V at Tr2 collector, 6.7 V at Tr2 emitter and 6.4 V at the base.

Clearly, then, Tr2 base is 0.3 V negative with respect to its emitter. Owing to the direct coupling, we have 4.4 V on Tr3 base relative to 4.5 V on its emitter. This means that the base is 0.1 V *negative* with respect to emitter. Tr3 is thus conducting a little (passing $I_c$).

However, due to R5, the voltage on Tr4 base is 4.6 V with respect to 4.5 V at its emitter. This means that the base is 0.1 V *positive* with respect to the emitter. Tr4 is thus also conducting a little.

R5 is the biasing resistor which is chosen in value to provide the correct biasing conditions. The circuit is actually arranged so that the voltage at point A with respect to chassis is slightly higher than half

Preliminary circuit and transistor tests    67

the battery voltage. Without this compensation the maximum positive excursions of the signal would be a little less than the negative ones owing to the voltage dropped across the emitter resistor R3 (Fig. 2.8) *plus* the bottoming voltage of Tr2.

Reverting back to the practical circuit in Fig. 2.8, we see that instead of the single bias resistor R5 (Fig. 2.9), a preset R7 and a parallel-connected thermistor are used. This network is for stabilisation. Without it the tendency for the quiescent current of the output pair to increase with rising temperature and decrease with falling temperature can result in bad distortion.

A low temperatures, especially, the output transistors tend to veer towards complete collector current cut-off without stabilisation in addition to that given by the 1-ohm emitter resistors, R2 and R3. The thermistor overcomes this trouble. As its temperature falls its value increases, thereby making Tr3 base go less positive and Tr4 base more positive. This action pushes up the $I_c$ of each output transistor and the stage as a whole is prevented from working too close to cut-off, a condition that can cause 'crossover distortion'. This is considered in Chapter 4.

In Fig. 2.8 further stabilisation is provided by the d.c. path between the junction of R2 R3 and the emitter of Tr1. Tr1, in fact, acts as a d.c. amplifier in this respect and compares the voltage at its base with the mid-point voltage at R2 R3 junction. The high loop gain of the amplifier thus fixes the mid-point voltage with respect to the base voltage of Tr1. This reduces the effects of transistor and component spreads.

It is interesting to note that the drive signal from the common-emitter Tr2 is applied between the base and emitter of both Tr3 and Tr4. This is because the drive signal across the collector load (R8) of Tr2 appears between these electrodes by reason of the a.c. coupling of C1.

**Push-pull effect**

Another aspect of interest with the complementary type of transistor audio amplifier is the lack of separate phase splitting. In Fig. 2.8 the speaker is connected capacitively, through C1, and, in spite of the fact that both output transistors are driven from the same signal point (i.e., collector of the driver Tr2), push-pull action arises because a positive half-cycle of drive causes conduction of the n-p-n transistor and cut-off of the p-n-p transistor, while a negative half cycle causes the p-n-p transistor to conduct and the n-p-n partner to cut off. More is told about the signal conditions of audio amplifiers in Chapter 4.

Analysis of directly coupled and complementary transistor stages is undertaken along the lines already discussed for single and capacitively coupled stages. The things to look for are the type of transistor (i.e., p-n-p or n-p-n), the polarity of voltages and the common electrode. For instance, the $V_c$ of Tr3 in Fig. 2.8 would be measured relative to the emitter. That is, at the junction of R2 R3. This method of measurement is obvious so far as Tr2 and Tr4 are concerned, of course, With Tr1, $V_c$ would be measured relative to the junction of R2 and R3, as with Tr3. In complex circuits, such as that in Fig. 2.8, network currents should be discovered from the voltage reading obtained across a resistor carrying the current. This will avoid the circuit from being disturbed due to the extra resistance of a current meter. Nevertheless, the loading effect of the voltmeter, even across relatively low value resistors, should always be borne in mind.

### The testmeter

A testmeter of not less than 10 000 ohms/volt should be employed, unless a higher resistance meter is available, such as 20 000 ohms/volt or even 100 000 ohms/volt. This is because it is often necessary to make voltage measurements in transistor equipment with a full-scale deflection of 1 V, which may be the lowest range of a multirange testmeter. The terminal resistance is then equal to the ohms/volt value, of course.

For speedy fault finding, the terminal resistance of the testmeter can often be used purposely to shunt a circuit resistor. Going back to Fig. 2.3, for instance, the prods of a voltmeter could be connected across R2 to give a small change in $V_{be}$, for the $I_c$ change test. Similarly across $R_b$ in circuit (a). In the latter case, however, the meter should be switched to a higher voltage range to avoid inciting too great an $I_b$. Remember that the terminal resistance of any voltmeter is equal to the ohms/volt value multiplied by the full-scale voltage range selected. A 20 000 ohms/volt meter has a terminal resistance of 200 k when set to the 10-volt full-scale range, while a 10 000 ohms/volt meter set, say, to 5-volts full-scale has a terminal resistance of 50 k.

When adopting this artifice, care should be taken not to connect the meter to a circuit across which the voltage exceeds the full-scale setting of the instrument. Also, it is best to apply the prods so that any indication on the meter is in a forward direction. If the needle backs against its stop, one cannot tell whether the movement is being overloaded.

## Use of ohmmeter

The ohmmeter of a multirange testmeter can also be used to advantage for passive testing. That is, with the equipment switched off. Even without disconnecting the transistor from circuit it is often possible to determine whether or not the collector and emitter junctions are in order.

The basic circuit of many ohmmeters is shown in Fig. 2.10. When the test terminals (prods) are shorted the multiplier resistor and battery are arranged so that the movement indicates approximately full-scale deflection. The scale is calibrated direct in 'ohms', with full-scale deflection corresponding to 'zero ohms'. The 'set zero' control allows the pointer to be set accurately to this point. When an unknown value resistor is connected across the terminals, with the short now removed, the current is reduced and the pointer deflects to something less than full-scale, depending upon the value of the test resistor. The $R$ value is then read direct from the scale.

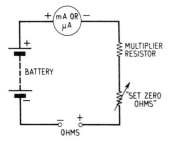

Fig. 2.10. Basic ohmmeter circuit. note that the terminals on a multirange testmeter which incorporates an ohmmeter may be polarised opposite to that shown here

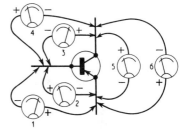

Fig. 2.11. Transistor tests that can be performed with the ohmmeter. With an n-p-n transistor polarities are reversed

Clearly, then, an ohmmeter connected across a transistor junction will give an indication in terms of forward current or reverse current, depending upon which way round the terminals or test prods are connected.

Fig. 2.11 shows the various combinations and polarities. Assuming that the ohmmeter terminals are marked positive and negative in accordance with Fig. 2.10 circuit (that is, with the actual polarities of the battery corresponding to the polarities marked on or against the terminals), then in Fig. 2.11 we have the following.

Test 1 showing forward current in the emitter junction, test 2 reverse current in the emitter junction, test 3 forward current in the collector junction, test 4 reverse current in the collector junction, test 5 showing a high resistance between collector and emitter and likewise with test 6.

Forward current, of course, is revealed by a low 'ohms' reading and reverse current by a high 'ohms' reading. It should be noted, however, that on some multimeters which can be switched to read 'ohms' the positive terminal may be connected to battery negative and the negative terminal to battery positive.

Test 6 measures $I_{ceo}$, of course, and here we would expect a very high 'ohms' reading. Test 6 is reproduced in Fig. 2.12(a). In Fig. 2.12(b) a 220 k resistor is added between the negative collector and the base. This will inject a small amount of forward current into the emitter/base junction from the battery in the ohmmeter, and, if the transistor is in good order, the high 'ohms' reading will change to a low one, as shown.

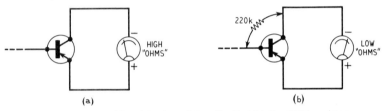

Fig. 2.12. Testing $I_{ceo}$ at (a) and the 'transistor effect' at (b). The 220 k resistor introduces a small forward current into the base

Junction tests can be made with the transistor in situ, as already mentioned. However, there will inevitably be shunt effects due to network resistors, and these must be taken into account. Similarly, if it is attempted to measure values of resistors connected to a transistor circuit, the forward and reverse conductions of the transistors must be taken into account.

A typical illustration is the in situ measurement of R1 in Fig. 2.8. Test A will shunt R1 by the forward resistance of Tr2 emitter junction and the reading will be somewhat below the true value of the resistor. Test B, on the other hand, results in the much higher reverse resistance of the junction appearing across the resistor and the reading obtained will be close to the true value of the resistor.

## Testing integrated circuits

In general it is best to regard an in-circuit i.c. in the same light as a transistor from the servicing point of view. Accurate assessment of the separate stages of an i.c. is virtually impossible under practical conditions. The first move should be to ensure that all the connections that should be in receipt of supply potential are, in fact, suitably energised. That is, to ensure that the supply feed circuits are not themselves responsible for any fault. It is noteworthy that a defect in a part of the i.c. can reflect incorrect voltages to some of the connections, particularly when these are connected to high source resistance values. Thus, if such a condition is observed in faulty equipment under test, whilst subsequent tests prove conclusively that the feed circuits themselves are not in any way responsible, then there would be a very good case to remove the i.c. for a special test or for substitution.

Removal of an i.c. is not a particularly simple matter. Care must, of course, be taken to prevent a high net flow of soldering iron heat to the device through the leadout wires or tags. By far the best plan is to use one of the specially developed desoldering tools, of which there are many varieties. To avoid undue overheating it sometimes pays to let the device cool off after each solder connection, and this is applicable whether removing a suspect device or fitting a replacement, for there is always the possibility of the suspect not being faulty after all, though as many tests as feasible should be performed prior to extraction to minimise such a possibility.

Dynamic testing is one of the best ways of proving whether an i.c. is working as it should in the overall design assuming that static voltage tests fail to point definitely to an i.c. defect. With some items of equipment it is possible to exploit the normal signal in the equipment and then to use an oscilloscope, with a low-capacitance probe if necessary, to check and monitor the signal at the input, output and intermediate terminations of the i.c. If it is impossible or difficult to use the normal in-circuit signal, then a signal from a test generator should be applied at points prior to the i.c. in the general manner of signal tracing. Indeed, the oscilloscope is now a very important testing 'tool', saving quite a lot of time and effort in servicing the latest generation of domestic-based electronics. Readers interested in this mode of testing may find my companion book *Servicing with the Oscilloscope* (Newnes Technical Books) helpful.

It is, of course, impossible in a short space to detail full i.c. testing procedures and the results to be expected, since these will obviously be related to the complexity and the nature of the i.c. and to the overall circuit in which it is employed. It is best to work in conjunction with the circuit diagram, and it helps one to determine the potentials and signals that should be present at the various leadout wires or tags if a circuit of the i.c. is also available.

## 72 Preliminary circuit and transistor tests

Stage bypassing with a capacitor can also speed fault diagnosis (see page 46).

Since the operating gain of an i.c. is determined by feedback, it follows that the gain (and the frequncy response) of an item of equipment in which an i.c. is employed can alter with change in value of a feedback component. Fig. 2.13 gives an example of the components involved in a feedback circuit, where $R_f$ is the feedback resistor and $R_s$ is the 'summing' resistor. In this case the output is equal to $-R_f e_{in}/R_s$. Sometimes a reactive circuit is included in the loop for response tailoring or equalisation, such as in the magnetic pickup preamplifier of a hi-fi amplifier.

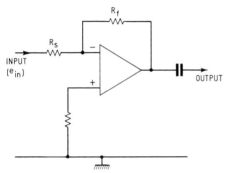

Fig. 2.13. Basic feedback in i.c. configuration.
There are other modes (Also see Fig. 1.37)

I.C.s have two inputs, one usually denoted (+), which is the non-inverting input, and the other denoted (−), which is the inverting input.

I.C.s are more difficult to test out of circuit than transistors and owing to the many different species currently at hand and the new ones becoming available a sort of general-purpose tester is barely feasible. However, it is possible to display the transfer characteristic of an i.c. on an oscilloscope and therefrom glean an idea of the open-loop gain, Fig. 2.14 shows the basic setup required for this.

Horizontal deflection of the trace is provided by the sawtooth which is proportional to the i.c. input voltage, while the output voltage of the i.c. provides the vertical deflection. The nature of the display is shown in Fig. 2.15, from whence the voltage gain is $y/x$.

With some oscilloscopes it is possible to scale down the timebase waveform and use this as the input signal. For this the oscilloscope should be equipped with a terminal delivering the timebase waveform.

Preliminary circuit and transistor tests 73

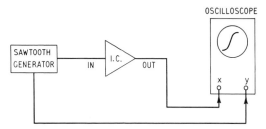

Fig. 2.14. Setup for i.c. transfer display (see text)

Fig. 2.15. Nature of transfer display, where gain is y/x

The i.c. will, of course, have to be connected to a suitably supply circuit and a network will have to be provided for offset balance. In some cases it is possible to test an i.c. in situ by this technique, using the equipment supply, after first suitably isolating the input and output circuits.

An oscilloscope is also useful for detecting spurious oscillation in an i.c.; but since the frequency may be very high an instrument with a wideband Y amplifier is essential. I.C. oscillation is also signified by abnormally high supply current which in a high resistance feed circuit could reduce the terminal voltages, this is another factor to bear in mind if voltage tests give unusual results!

## Testing f.e.t.s

F.E.T.s, and the insulated gate variety in particular, are vulnerable to gate breakdown due to transients and static effects. Some devices are equipped with a shorting link or ring between the gate and the other electrode leads to avoid possible damage due to static changes, etc. It is desirable for this 'short' to remain in position until after the device has been soldered into the circuit, for even transients from a soldering iron can sometimes lead to failure.

Current and voltage tests on the source and gate electrodes are generally permissible, and the techniques involved are not much different from those detailed for bipolar transistors.

This chapter is summarised in Fault Diagnosis Summary Chart 1.

**Fault Diagnosis Summary Chart 1: Basic Circuit Test**

| Condition | Probable Cause | Check |
|---|---|---|
| Zero $V_c$ | (i) Faulty power supply | (i) Battery or mains unit, feed resistors and switch |
| Zero $I_c$ | (i) Open-circuit collector load<br>(ii) Open-circuit collector junction<br>(iii) Zero $I_b$ | (i) Load resistor or coil<br>(ii) Transistor<br>(iii) $I_b$ and $V_{be}$ |
| High $V_c$ | (i) Zero $I_c$<br>(ii) Zero $I_b$<br>(iii) Zero $I_c$ | (i) Transistor<br>(ii) $I_b$ and $V_{be}$<br>(iii) $I_e$ and emitter resistor |
| Low $V_c$ | (i) Very high $I_c$<br>(ii) Collector load gone high in value<br>(iii) Low power supply voltage<br>(iv) Leak in coupling capacitor | (i) Collector junction for short<br>(ii) Collector load<br>(iii) Power unit or battery<br>(iv) Coupler for insulation |
| No $I_c$ change with $I_b$ change | (i) Collector junction short<br>(ii) Incorrect base bias | (i) Transistor<br>(ii) Biasing |
| High $I_e$ no $I_c$ | (i) Emitter junction short | (i) Transistor |
| Zero $I_e$ high $I_c$ | (i) Collector junction short | (i) Transistor |
| Zero $I_b$ zero $I_c$ | (i) Zero $V_{be}$ | (i) Biasing on base |
| High $I_b$ | (i) Short in emitter junction | (i) Transistor |
| Distortion | (i) Transistor bottoming (high $I_{ceo}$)<br>(ii) Transistor clipping (low $I_b$)<br>(iii) Leaky coupling capacitor | (i) Base bias, poor d.c. stabilisation, excessive temperature<br>(ii) Base bias<br>(iii) Coupling capacitor |
| Noisy volume control | (i) Leaky coupling capacitor<br>(ii) Faulty control | (i) Coupling capacitor<br>(ii) Control |

# 3 Signal conditions and tests

In Chapter 2 we were mostly concerned with tests of direct voltages and currents in transistor circuits. These, we saw, set the operating conditions of the transistor so that the alternating signal voltages and currents injected into and taken from the circuit can be handled with optimum advanage with respect to the required application.

As a bipolar transistor is a current-operated device, it must receive essentially signal current. In the case of a simple, common-emitter amplifier, the signal current to be amplified is injected into the transistor in series with the base current. From first principles, this means that the resulting signal current in the collector is likely to be beta times the signal current in the base circuit.

On positive swings, the signal current pulls down the collector current from it d.c. standing value in p-n-p transistors, while on negative swings it pushes the collector current up above its d.c. standing value. The converse is true, of course, with n-p-n transistors.

The effect is similar to that of decreasing and increasing the base current. Signalwise, this is accomplished automatically by the applied signal current waveform. The standing collector current is established by the d.c. conditions, as we have seen, and it will now be appreciated that for the signal to be handled correctly the standing $I_c$ must be optimised to suit the signal conditions and requirements.

## Bottoming and clipping

Let us suppose that the base bias is too small, resulting in a very small value of $I_c$. Then when a symmetrical current waveform of signal is applied to a p-n-p transistor the positive peaks will drive the transistor towards collector current cut-off, while the negative peaks will normally operate well within the signal limits of the transistor. If the input signal is of full-drive amplitude, the transistor would, in fact, cut-off on the positive peaks, resulting in bad distortion.

On the other hand, if the base bias is too large, then the negative peaks of input signal would drive the transistor towards saturation, while the positive peaks of signal would normally be handled correctly. This effect is called *bottoming*, and it occurs completely when an increase in base current fails to influence the collector current.

Clearly, then, the d.c. operating conditions are arranged to ensure that the applied signal is handled on both half cycles within the operating limits of the transistor. A change in value of a component which might cause a change in the d.c. operating conditions would be likely to cause distortion, as we shall see in later chapters.

We have seen that the $I_c$ is made up both of collector current resulting from the normal transistor effect and of collector leakage current. Should the transistor be subjected to abnormally high temperatures, the leakage current would rise and the $I_c$ may go above its normal standing value. A stage under this condition handling a signal would tend to bottom at a smaller input signal amplitude than the same stage operating under normal temperature conditions.

Conversely, collector current cut-off conditions may arise due to an excessively low temperature. These effects are sometimes troublesome in transistor radio sets when operating under abnormally high temperature and abnormally low temperature environments. Artifices are available, however, to minimise the effects and to stabilise the operating conditions over a wide range of temperatures. These are considered in Chapter 4.

## Basic factors of stage gain

Let us now investigate the common-emitter circuit in Fig. 3.1 from the signal point of view. Although shown with a p-n-p transistor, the same principles apply with n-p-n devices. Base bias is established by the base potential divider R1/R2. Let us suppose that this is set for optimum signal swing of both positive and negative half cycles of the applied signal and that the signal amplitude neither bottoms nor cuts off the transistor.

This represents the normal class A operating conditions. The action is that a swing of a few microamperes of base current gives rise

Signal conditions and tests 77

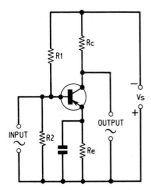

Fig. 3.1. Basic common-emitter stage. This is used in the test for describing the signal conditions

to a swing in the order of milliamperes of collector current. This means, when the design is for voltage gain, that the voltage swing across the collector load $R_c$ will be significantly greater than the voltage swing across the base resistor R2. In this manner the transistor amplifies the signal voltage. However, factors other than the resistors themselves come into calculations of signal gain, and the calculations must also take account of the circuit configuration (i.e., common-emitter, common-base or common-collector, see Figs. 1.9 and 1.13). From first principles, voltage gain is a function of the signal current gain of the transistor ($\alpha'$) (see page 22), the collector load $R_L$ and the output impedance $Z_{out}$ in the manner of $\alpha' R_L/(R_L + Z_{out})$. There is more to it than this, however, since $Z_{out}$ is itself a complex function tied up with the nature of the transistor; and other factors such as transistor input impedance and the effect that this has on the output impedance need to be included to secure an accurate appraisal of voltage gain (see also under *Signal Effects* on page 79).

**Table 3.1**

|  | Common base | Common emitter | Common collector |
|---|---|---|---|
| Current gain | ~1 | High | High |
| Voltage gain | High | High | ~1 |
| Input impedance | Low | Medium | High |
| Output impedance | High | Medium | Low |
| Power gain | Medium | High | Low |
| Cut-off frequency | High | Low | Depends on $R_L$ |
| Voltage phase shift at low frequencies | ~Zero | ~180° | ~Zero |

The emitter circuit and the coupling to a following stage also influence the efective gain.

Current gain in a signal-carrying circuit is also similarly interrelated with factors other than $\alpha'$, and full details of such calculations can be found in the *Mullard Reference Manual of Transistor Circuits*, for example.

Power gain is considered as the product of the current gain and the voltage gain, and under matched conditions this is highest in the common-emitter mode and least in the common-collector mode.

Table 3.1 compares the operational parameters.

Amplifier sensitivity was sometimes given in terms of input current rather than input voltage. An early Mullard 10 watt class AB power amplifier, for instance, has a sensitivity parameter of 140 μA for 10 watts output.

Now to get back to the circuit in Fig. 3.1. The standing value of $I_c$, sometimes called the *static value*, is revealed on the output characteristics (see Fig. 1.19) by means of a *load line*. This is originally drawn on the characteristics by taking into consideration the value of $R_c$ and the supply voltage.

If it is supposed that a transistor is driven to the limits of cut-off on the one hand and saturation or bottoming on the other, then when it is driven to cut-off the voltage between the collector and emitter ($V_{ce}$) will be equal to the supply voltage ($V_s$). This is because no current flows through the transistor (actually, of course, leakage current flows but this is normally negligible).

When driven to saturation, the transistor is approaching a short-circuit between the collector and emitter. If we ignore the effect of the emitter resistor $R_e$, we can see that almost all $V_s$ will appear across $R_c$. The current in $R_c$ will then be equal to $V_s/R_c$ (Ohm's law).

Now, if we study the load line on the output characteristics, we shall see that the line is terminated at the $V_s/R_c$ point on the $I_c$ axis and at the $V_s$ point on the $V_c$ axis. A load line drawn between two such points is shown on the characteristics in Fig. 3.2.

A load line is used to fix the working point to allow the required swing under drive conditions within the limits of saturation and cut-off. In Fig. 3.2 it will be seen that the working point to satisfy these conditions is established with a base current ($I_b$) of 30 μA. This produces a little over 3 mA of collector current ($I_c$) through the load $R_c$ (in this example, $R_c = V_s/7$ mA, which is 12/7 kilohms, or very slightly below 1714 ohms).

Thus, resulting from a swing of $I_b$, the $I_c$ will swing either side of its static value and the $V_c$ will also swing either side of its static value up and down the load line.

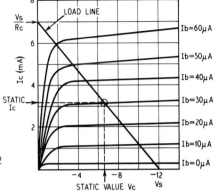

Fig. 3.2. This load line on the output characteristics is explained in the text. Note that $V_s/R_c$ in this example is equal to 7 mA. Thus, with a $V_s$ of 12 volts we have a collector load $R_c$ equal approximately to 1 700 ohms

While we are not interested essentially in the design of transistor circuits in this book, it is rather important that we have a reasonable understanding of how the working point is fixed by the designer and why it is so fixed, for a fault condition may well shift the working point from its correct value, so it is as well that we know what effect this is likely to have on the performance of the circuit – and why.

To sum up, then, the working point is established in the beginning by the $I_b$ being set to give the required static $I_c$ in relation to the collector load resistor and the supply voltage. Clearly, a shift in any of these factors would also shift the originally intended working point.

The transistor itself also influences the working point. For instance, should the $I_c$ rise due to an increase in temperature or due to spreads in the characteristics of a replacement transistor, the working point would move up the load line. Conversely, transistor-spreads in the opposite sense or a drop in $I_c$ would cause the working point to move down the load line. These factors have already been mentioned, and they are investigated more from the practical aspect in Chapter 4.

## Signal effects

When a transistor is subjected to signal voltages, the effects of reactances and impedances come into play. These influence the stage gain, as also, of course, does the type of transistor. We saw in Chapter 1 the requirements for v.h.f. and u.h.f. transistors. The frequency limitations of transistors are now generally specified by the term *transition frequency*, or $f_T$. This is the frequency at which beta falls to unity. This parameter is sometimes called the *gain-bandwidth product*.

From the practical aspect, there is little point in replacing, say, a.u.h.f. transistor with a device whose $f_T$ is little above 100 MHz, since the stage gain at u.h.f. would then be well below unity! Conversely, if a device with a high $f_T$ is used in a circuit which was originally designed for a transistor with an $f_T$ of, say, 20 MHz or less, then one might have difficulty in maintaining the stability of the circuit under certain conditions.

## Main *h* parameters and equivalents

Signalwise, there are four primary parameters frequently encountered in the literature. Although we shall rarely adopt these in our day-to-day fault-finding activities, it is nevertheless as well to know what they are.

The first is *the input impedance for a constant output voltage* (i.e., the slope of the input characteristics under signal conditions). This is defined as a small value of input voltage divided by the resulting small value of input current. It is given the symbol $h_{11}$. The small letter $h$ refers to the so-called hybrid or $h$ parameters.

The second is *the current amplification factor for a constant output voltage* (i.e., the slope of the transfer characteristic). This is defined as the small value of signal output current divided by the small value of input current to cause it. This, as we have already seen, is also termed beta, etc. It is given the symbol $h_{21}$.

The third is *the reciprocal of the output impedance*. That is, the output admittance for a constant input current (i.e., the slope of the output characteristic). This is defined as a small output current divided by a small output voltage. It is given the symbol $h_{22}$.

The fourth is *the voltage feedback ratio for a constant input current* (i.e., the slope of the feedback characteristic). This is defined as a small input voltage divided by a small output voltage, the former being produced by the latter. It is given the symbol $h_{12}$.

Thus we have $h_{11}$, $h_{21}$, $h_{22}$ and $h_{12}$. Instead of these symbols, equivalent symbols of $h_i$, $h_f$, $h_o$ and $h_r$ respectively may be used. Here the subscripts i, f, o and r refer to input impedance, forward conduction, output impedance and reverse conduction respectively. Note here that $h_f$ is the symbol of current gain. In the common-emitter mode, a further subscript e is added, giving $h_{fe}$. This, then, is the small signal current gain equivalent to *beta, alpha dash* and $h_{21}$. The capital letters FE are sometimes used as subscripts to indicate the current gain at d.c. (i.e., d.c. beta). Thus we get $h_{FE}$.

In an endeavour to put some stability into this confusion of terms which has grown up with the evolution of the transistor, we would summarise by stating that the common symbols for current gain at a.c. (i.e., small signals) are beta ($\beta$), alpha dash ($\alpha'$), $h_{fe}$ and $h_{fe}$, and that the common symbols for current gain at d.c. are beta, beta with a bar ($\bar{\beta}$), beta d.c. ($\beta_{d.c}$), alpha nought dash ($\alpha_0'$) and $h_{FE}$. Unfortunately, these terms are not consistently employed.

For the sake of the record, it should be noted that there are two additional types of transistor parameters. These are called $z$ parameters (also modified $z$ parameters) and $y$ parameters. They are employed essentially for network analysis and design and would not normally be used in fault-finding and repair work.

There is one aspect of the transistor which should be particularly noted. This is related to the $h_r$ or $h_{12}$ parameter, which is the parameter of reverse conduction. This implies that a signal at the collector of a transistor, due to internal feedback, can get back to the input. The amount of signal being reflected back in this way is governed by $h_{12}$.

To those readers conversant with the action of thermionic valves and their circuits, it should be mentioned that there is no equivalent effect in a valve. This internal feedback is one of the sole peculiarities of the transistor. It means that the input impedance of a transistor can be affected by the conditions at the output load.

## Types of signal

We now come on to signal tests in transistor stages. In this connection we shall be dealing with audio, video, radio-frequency and intermediate-frequency signals, symbolised a.f., v.f., r.f. and i.f. respecitvely. The r.f. and i.f. signals will be both unmodulated and modulated and will range from low-frequency (l.f. 30 to 300 kHz), through medium-frequency (m.f. 300 kHz to 3 MHz), high-frequency (h.f. 3 to 30 MHz), very high frequency (v.h.f. 30 to 300 MHz) to ultra high frequency (u.h.f. 300 to 3 000 MHz).

A.f. signals range from d.c. to about 20 kHz and v.f. signals from d.c. to about 6 MHz. Some signals will be of pure sine wave makeup and others of pulsed, square, triangular, transient and etc. form, and the signal equipment will be concerned with both generation and amplification.

The ordinary transistor radio receiver, for instance, embodies r.f., i.f. and a.f. amplifying circuits and a sine wave generator at r.f. for the local oscillator signal. The television set, in addition, contains sawtooth (i.e., square and triangular wave) generators in the time-base circuits and pulse networks in the synchronising circuits. The video signal itself can also be considered as pulses of square/triangular form and of a transient nature. In the transistor tape recorder, we find a.f. amplifiers along with a sine wave oscillator delivering an l.f. signal for record bias (note that in tape recorders a sinewave from, perhaps, 50 kHz to 200 kHz, depending on machine, is used for bias and erase – usually called h.f. bias/erase).

Radio and television receivers, of course, feature also detector circuits for extracting the modulation signal applied during the transmitting process to the r.f. carrier wave. In transistorised receivers, the detector usually takes the form of a germanium or silicon diode.

Thus to test signal circuits, we must first find out whether they are concerned with signal generation or amplification, and the range of frequencies over which they are designed to operate. Oscillator and signal generating circuits are dealt with in Chapter 6.

## Signal injection

Fault finding in transistor amplifier circuits is facilitated by the fact that the equipment of which the amplifier under test is a part usually embodies some kind of detector or indicator. This may be a tuning indicator, level- or monitor-meter, loudspeaker, headphones or earpiece, relay or cathode-ray tube.

This means that it is usually possible to inject a test signal of suitable frequency into the input of the equipment and employ the inbuilt detector or indicator to announce its presence at the output. Should the applied test signal fail to be indicated in this way, and we are sure that the detector or indicator itself is in order, then we would be pretty certain that the fault lies somewhere in the stage or stages between the point of signal injection and the detector or indicator.

A simple illustration is a two-stage audio amplifier which is completely dead. Here an a.f. test signal would of course be used. This would be applied first across the loudspeaker and its level adjusted until it is heard as a tone from the indicator (loudspeaker, in this case). This would prove that the indicator itself is working correctly.

The signal would be moved from the loudspeaker to the input of the second amplifier (i.e., that feeding the loudspeaker). If this stage is working properly, a signal of far lower level than that applied direct to the loudspeaker would give a tone of equal volume to that originally obtained. lack of response would tell that the second amplifier is defective. A normal response with the signal applied to the input of the second amplifier would lead to the signal being transferred to the input of the first amplifier. A correctly working first stage would call for a much lower level of test signal to provide the original volume of tone from the loudspeaker. Should this, in fact, happen, the trouble would lie between the input of the network where the signal is actually applied and the input electrode of the first stage transistor – i.e., the base in the common-emitter mode.

Fig. 3.3. For signal injection testing, the test signal is applied in turn to the input of the various sections and stages, starting from the loudspeaker, as this diagram shows

The block diagram in Fig. 3.3 shows the various test points for signal injection relative to a two-stage audio amplifier. Exactly the same techniques are adopted no matter how complex the equipment or the type of detector or indicator embodies within the equipment. With the vision channel of a television set, for example, the cathode-ray tube would be used as the indicator. The injected video-frequency signal would produce bars on the screen, as audio produces tone from a loudspeaker. More is said testing in television circuits in Chapter 4.

## Signal tests

In straight amplifiers where no indicating embodiment exists, the equipment under test must either be connected to an instrument for detecting the applied test signal or to a complete amplifier system or receiver, depending upon the type of equipment under signal test.

Let us take the case of a u.h.f. aerial amplifier for television. Such a device on its own would demand for testing a generator of u.h.f. signals and a detector of u.h.f. signals. The former could be an ordinary signal generator (note here that where a u.h.f. generator is not available, the third or fourth harmonic from a v.h.f. signal generator tuned to the top band can be employed, or even the u.h.f. transmission itself). The latter could be a u.h.f. signal strength meter.

The test procedure would be first to measure the strength of the u.h.f. signal applied to the input of the amplifier and then to measure the strength of the signal at the output of the amplifier, as shown in Fig. 3.4.

If a television set and signal from the aerial are employed, one may expect the amplifier connected between the aerial signal and the aerial socket on the set to boost the reception. In practice, a distinct boost effect of sound or vision may not be evidenced. This is because

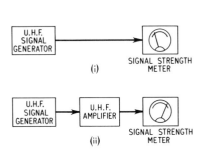

Fig. 3.4. The simplicity of testing a u.h.f. amplifier. First, the generator signal is applied direct to a signal strength meter and the reading (i) noted, and secondly the generator signal is applied to the signal strength meter via the amplifier and the reading (ii) again noted. The difference between reading (ii) and reading (i) is the gain of the amplifier. Note, however, the importance of ensuring that the amplifier, signal generator and signal strength meter impedances match

the television set incorporates automatic gain control (a.g.c.) in both the vision and sound channels. Nevertheless, the u.h.f. amplifier should have a relatively low-noise specification, meaning that its inclusion should tend to reduce the amount of 'snow' on vision and 'hiss' on sound.

V.h.f. aerial amplifiers would be tested for signal in a like manner. A.f. preamplifiers, such as those used to boost a pickup signal or the output of a microphone, could be tested either in conjunction with an audio amplifier or in isolation with instruments. This time, of course, the applied signal would be within the audio range of the amplifier and the detector would be designed to respond to audio signal. A typical audio detector is a cathode-ray oscilloscope or valve voltmeter.

The various types of amplifying stages in transistor radio and television receivers can also be tested by signal injection, ensuring that the frequency and level of the signal injected are suitable for the stage under test in any particular section. More practical information on signal testing is given in Chapters 4, 5 and 7.

## Signal tracing

Another method of testing designed for speedy fault diagnosis is termed signal tracing. This technique employs an instrument which is

## Signal conditions and tests

itself a detector of signals. It is called a signal tracer. In its simplest form it may be a headphone set or earpiece.

The idea is to trace by sampling the signal between the input and output of the equipment at various points. A two-stage audio amplifier, for example, would be tested in this way by connecting the detector first to the programme source (say the output of a pickup), then to the input of the amplifier, to the output of the first stage, to the input of the second stage and finally to the output of the second stage.

The signal would in this way be traced right from the pickup to the speaker, and the point of discontinuity in the channel would be revealed by lack of response in the detector or signal tracer. Let us suppose that the pickup signal is heard at the output of the first amplifier stage (see Fig. 3.5) but not at the output of the second amplifier stage. We would be sure that the fault exists somewhere between the output of the first stage and the output of the second stage. By sampling the signal at the input of the second stage we could discover whether the trouble lies in the interstage coupling network.

Fig. 3.5. For signal tracing tracing, the 'signal tracer' is used to monitor the signal from the start to the finish of the amplifier channel, as this diagram shows

Signal tracers also include diode detector probes, enabling a modulated r.f. or i.f. signal to be heard from headphones or loudspeaker. The detector, of course, simply extracts the modulation of the signal and applies this to an inbuilt amplifier, in the same way as the test or sampled audio signal is applied.

There are other ways of testing for signal response as a means of speeding fault finding in transistor equipment. These are considered in the subsequent chapters of this book.

## Impedance considerations

When signal is applied and extracted from amplifier sections in any piece of transistor equipment, special consideration must be given to the input and output impedance so far as the signal test is concerned. Confusing results may occur if it is attempted to inject or extract signal into or from a low impedance circuit by the use of a generator or detector with a high impedance termination. Conversely, trouble may well be encountered by the use of a generator or detector with a low impedance termination in conjunction with a high impedance circuit.

An example in this respect lies in the testing of a v.h.f. or u.h.f. aerial amplifier or booster. Such a device generally features unbalanced (i.e., coaxial) input and output circuits of 75 ohms impedance (sometimes 300 ohms balanced in countries other than Great Britain).

This means, then, that both the generator or aerial and the receiver or signal strength meter must also be terminated at about 75 ohms, unbalanced. These conditions are easy to satisfy, since British signal generators, aerials, aerial downleads, signal strength meters and television sets are all terminated at around 75 ohms.

Should the signal source be terminated at, say, 2 000 ohms, then the input of the amplifier would be incorrectly loaded, a condition which may well induce instability and unpredictable meter indications. Likewise, the output of the amplifier would be incorrectly loaded if the input of the receiver or signal strength meter had an impedance differing widely from that of the amplifier output.

The input channels of audio amplifiers and control units are also designed for specific impedance circuits. A crystal or ceramic pickup channel, for example, would probably have an input impedance of 2 megohms, while the impedance of a magnetic pickup channel may only be a few hundred ohms, depending on type. Radio input channels range up to about 100 000 ohms impedance, while microphone channels are often at a much lower impedance, from 15 to about 600 ohms.

Audio amplifier outputs are designed for both medium and low impedance. A medium impedance (about 1 000 ohns) circuit might be found in the control unit of hi-fi amplifiers and tape recorders to enable the programme signal to be monitored or applied to the high input impedance of a second power amplifier or control unit or into the 'radio' input impedance of a second power amplifier or control unit or into the 'radio' input of a tape recorder.

The loudspeaker is always fed at low impedance, and in transistor equipment this may range from a few ohms up to several hundred ohms, depending upon the type and design of the output stage. Owing to negative feedback and the intrinsic low output impedance, audio power amplifiers are often regarded as constant voltage sources. Care must, therefore, be taken to ensure that the value of the output load connected is not abnormally low as to cause a very high audio current to flow, which could damage the output transistors unless somehow protected electronically or by fuses, though the latter may not act quickly enough.

## Coupling

Coupling is achieved either by ensuring that the instrument or ancillary equipment tied to the equipment under test suits the input or output impedance, as the case may be, or that the instrument or ancillary equipment has connected in parallel or series with it a resistance or impedance which will simulate the 'normal' loading requirements. Subsequent chapters dealing with fault tracing give details of such requirements.

It should be noted, however, that when a resistance or impedance is connected in parallel or series with the source impedance (i.e., the impedance into which the amplifier looks), and when the combination is applied so as to match the impedance of the amplifier, signal power is lost in the matching artifice.

This loss must be taken into account, of course, when absolute gain measurements are being performed, and when it is desired to determine the level of signal actually applied to an amplifier in conjunction with the calibrated attenuator of a signal generator.

The level of signal present across the input termination of the amplifier under test is as indicated by the attenuator of the signal generator only when the generator is correctly loaded, or used in accordance with the makers instructions. At different loadings, which may be demanded to suit the equipment to which the signal is being applied, the signal level will differ from that of the correctly loaded level by an amount dependent upon the extent of the generator loading.

Sometimes it may be necessary to employ an impedance matching pad between a signal generator and the input of an amplifier. While such a pad ensures that both the generator and the amplifier are loaded by their correct and individual terminal impedances, the pad itself is, in fact, an attenuator, and the amount by which it reduces the generator signal at the amplifier input must be taken into account. It should be remembered that signal loss is the price that has to be paid for correct matching. Moreover, maximum signal transfer from one network to another occurs only when the impedances of the connected networks are matched. Apart from impairing the signal transference, incorrect input and output matching in transistor equipment can grossly interfere with the performance and set up conditions of instability. A bad mismatch at the output of an amplifier could also result in a mismatch at the input due to the reverse conduction ($h_{12}$), as mentioned earlier.

It is generally permissible, however, to feed into a high input impedance of an audio amplifier, for example, from a low impedance source without incurring circuit displeasure. In this case the input is effectively 'seeing' a constant voltage source.

## Internal impedances

When applying and extracting signal to and from an indivdual stage in a piece of transistor equipment, the impedance at the point of application and extraction should be considered to avoid disturbing the normal operating conditions of the circuit as little as possible. The impedances at the input and output of a transistor single-stage amplifier are governed to a large extent by the configuration of the stage. These have already been briefly mentioned in Chapter 1. However, they are brought together again in Fig. 3.6 for the sake of convenience.

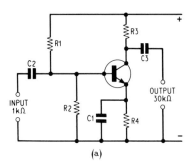
(a)

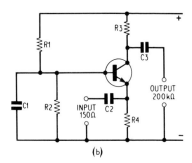
(b)

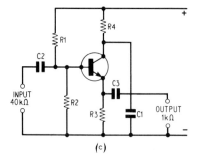
(c)

Fig. 3.6. The three transistor modes: (a) common-emitter, (b) common-base and (c) common-collector. Capacitor C1 in each circuit makes the appropriate electrode common to both the input and output signals. The power gain of (a) is in the order of tens of thousands of times, of (b) thousands of times, of (b) thousands of times and of (c) hundreds of times. Note the approximate input and output impedances indicated on the circuits

Here at (a) is the common-emitter mode, with the input signal applied to the base and the output signal taken from the collector. This mode has a medium to low input impedance and a medium to high output impedance. At (b) is the common-base mode. The input impedance of this mode is low, at the emitter, and the output impedance high, at the collector. At (c) with the signal applied to the base and taken from the emitter, giving the common-collector mode, the input impedance is medium to high and the output impedance low.

In all modes, R1/R2 forms the base potential divider, setting the base current, C1 is the capacitor which 'earths' the common electrode from the signal aspect, C2 is the input coupling capacitor, R3 is the load resistor and C3 is the output coupling capacitor. In (a) and (b) R4 is the emitter resistor. Its chief application in (a) is for d.c. stabilisation, as we have already considered, and in (b) as a load across which the input signal is applied. R4 in (c) is mainly for decoupling purposes. In some common-collector circuits the collector is taken direct to the supply line.

Note that in all cases the input signal is injected into the emitter/base junction, either at the base or at the emitter depending upon the mode.

## Gain

The most often used circuit is the common-emitter, because this mode provides the highest power gain, of the order of tens of thousands of times. The common-base mode can give power gains in the order of thousands of times, while the power gain of the common-collector mode is measured in hundreds of times.

The voltage gain is about equal in the common-emitter and common-base modes, but the difference is power gain between these two modes is due to the common-emitter circuit having a current gain well above unity (i.e., $h_{fe}$), while the current gain in the common-base mode is below unity (i.e., $h_{fb}$). While the common-collector mode has a current gain about the same as that of the common-emitter mode, the voltage gain of the common-collector circuit is only unity, which is the reason why the power gain of this configuration is below that of the others.

Single amplifier units are connected in cascade in the equipment as a whole in such a way that each stage is designed to supply the signal drive required by the following stage. The intercoupling networks may also take into account the difference between the output

## Signal coupling

When an instrument concerned with the signal is connected internally to an item of transistor equipment, care should be taken as far as possible to ensure that the circuit is not heavily loaded by the instrument. This is particularly important when stage gain measurements are attempted. As an example, the connection of a signal from, say, a 75-ohm source to the input of a transistor stage with an impedance of 5 000 ohms would badly swamp the stage input and considerably modify the performance of the circuit.

One way of connecting the signal would be through a 5 000-ohm resistor, with a suitable load resistor across the generator, as shown at (a) in Fig. 3.7. Note, however, that the series resistor in conjunction with the 75-ohm source then acts as an attenuator, and only a small fraction of the signal present across the source appears across the 5 000-ohm input of the stage.

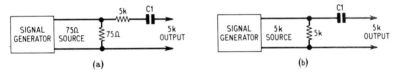

Fig. 3.8. Simple matching arrangements. A low impedance generator source can be raised for application to a medium/high transistor input impedance by the use of a series resistor, as shown at (a). A high impedance generator source can often be connected direct to a lower impedance input circuit, as shown at (b). Here the generator need not always be loaded by a resistance equal to its source (as shown), owing to the swamping effect of the circuit load. When feeding a circuit of higher impedance than its source, however, the generator itself needs correct loading, as at (a). Capacitor C1 in each circuit is for d.c. isolation and transient protection

Provided the signal generator is capable of delivering a signal of adequate level without overloading to outweigh the attenuation factor, this method of applying signal to a load of higher impedance than the source is perfectly satisfactory.

When the source impedance is significantly greater than that of the circuit which is being tested, there is not usually any need to step-down the impedance, for the circuit concerned is then only very little disturbed by the source impedance. However, it may be necessary to 'load' the source with its correct value, depending upon its nature, as shown in Fig. 3.7(b).

For simple 'go-no-go' tests and basic signal injection tests one need not worry too much about impedances, provided one understands the effects that the instrument loading can have on the circuits under test. Neverthless, when injecting an a.f. signal, a series resistance of about 10 000 ohms should be employed. This is to avoid distortion of the input current waveform by the non-linear input characteristics of the transistor. This has already been considered, and we must always remember that since a transistor is a current-operated device, it is the current signal waveform that is important.

It is also very important to remember that instruments or any equipment should never be connected to transistor circuit in such a way as to disturb the d.c. conditions. From the aspect of signal generator and the like, this means that coupling should be through a capacitor or inductive loop.

When mains-powered instruments and equipment are connected to a transistor circuit there is always the danger of the transistor being destroyed by the production of current or voltage transients. These can also be minimised by connecting the equipment through the lowest possible value capacitor. More is said about this very important subject of transient damage in Chapter 8.

# 4 Fault-finding in audio and video circuits

There are various other aspects of transistor circuits which we have not considered. However, as many of these are design aspects, as distinct from those of fault-finding, they will be referred to as we proceed only when they have a bearing on fault-finding.

## Audio amplifier arrangements

The basic audio amplifier comprises two sections if its purpose is to work a loudspeaker or other type of power transducer. The first section is the preamplifier which increases the level of the input signal sufficiently to drive the output section. The second, output section of the amplifier is designed to convert the signal to power to drive the transducer (e.g. loudspeaker).

Both sections often feature various stages. For instance, the preamplifier of an audio amplifier may have two transistor stages forming the input amplifier (to which the programme signals are fed), followed by a third amplifier stage designed round a tone control system. The input amplifier is designed for low-noise operation by the use of low-noise transistors and by running these transistors at a very low $I_c$. This part of the circuit may also incorporate equalisation networks for the programme signals, such as is needed for disc pickup and tape signals.

The final preamplifier stage facilitates variable tone control while also lifting the equalised and tone-controlled signals to a level suitable for driving the second section of the overall amplifier system, also at the required impedance. The final stage of the first section, then, can be considered as a 'buffer' stage to the second section.

The second section of the overall amplifier system invariably adopts a pair of power transistors in a push-pull stage. These are then driven by a phase-splitter or 'driver' stage. Thus, the second section usually incorporates two stages, the driver stage and the push-pull output stage.

An audio amplifier which is not required to produce power to work a speaker simply comprises one or a number of stages designed to give the required gain and interface impedances. For instance, a common-emitter amplifier alone can give quite a high gain at a low to medium input impedance and medium to high output impedance. By feeding the output of such a stage into a common-collector circuit, a low output impedance is obtained. A low input impedance and high output impedance stems from the use of a common-base circuit (also see the buffer amplifier circuit in Fig. 1.20).

### Inter-stage losses

We have already discussed the impedance aspect of transistor circuits in Chapter 3, but it must be emphasised that high multiples of gain are achieved not only by the cascading of separate amplifier stages but by ensuring good impedance matching between the stages. The use of resistive pads and networks for interstage matching or for matching signal sources can only result in attenuation. Thus, while the transistor stage itself may amplify, the stage or signal coupling may attenuate, thereby giving an overall unity gain characteristic. Designers often have to lose gain in couplings and restore it again by transistors to satisfy impedance requirements. This factor must be remembered while diagnosing for faults.

### Video-frequency amplifiers

Video amplifiers, such as those used to operate a picture tube in a transistor television set or to amplify the video signal from a camera TV tube, are concerned essentially with the amplification of small video signal currents and voltages. Video power is rarely needed.

Thus, a video amplifier is akin to a specialised audio preamplifier, but while the latter may respond only from about 30 Hz to 20 kHz, the former often has to give almost equal amplification from zero frequency (i.e., d.c.) to 6 MHz. The greater frequency range of video amplifiers is accomplished by special techniques, by the use of 'peaking coils' and by high-frequency transistors. From the fault-finding aspect, there is not a great deal of difference between the two types of amplifier, which is the reason why they are dealt with together in this chapter. But first let us consider the audio amplifier.

## Audio circuit faults

Fig. 4.1 is the circuit of a 10 W amplifier designed for 16-ohm speaker loading. It is the type that might be found in a radio receiver or sound channel of a high quality TV receiver. For stereo, of course, there would be two such amplifiers.

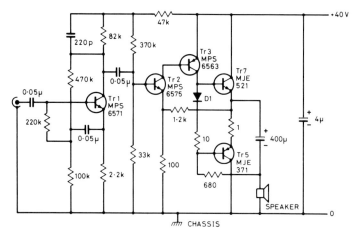

Fig. 4.1. Circuit of 10 W audio amplifier which has a high input impedance provided by bootstrapped Tr1 and an output suitable for driving a 16ohm speaker (see text)

Input signal is fed to Tr1 base, the transistor being biased by a potential divider and arranged as a low-level common-emitter amplifier. High input impedance is provided by bootstrapping (see later). Amplified signal swinging across the 82k ohm collector load is capacitively coupled to the cascaded pair Tr2/Tr3 which are in common-emitter mode. You will see that there is direct coupling here as well as between Tr3 and the bases of the push-pull pair Tr4/Tr5. These transistors are of complementary polarity (e.g., n-p-n and p-n-p). The correct, push-pull drive thus obtains without the need for phase splitting. The circuit has some of the features of those already discussed.

## Complete failure

Once it has been proved by the methods so far expounded that signal is reaching the input but failing to appear at the output, the first move would be to establish that the supply voltage is present, followed by a check of the d.c. operating conditions of the transistors. As already noted, directly coupled stages are stabilised by overall d.c. feedback, provided in Fig. 4.1 by the 1.2k ohm resistor from Tr4/Tr5 emitters back to Tr2 emitter. This transistor has near unity gain and works as a kind of 'comparator' so that the voltage at Tr4/Tr5 emitters holds at half the supply voltage (e.g., at the point where the speaker is connected).

The d.c. tests should soon reveal a significant abnormality, but if you are still in the dark apply an audio input signal and use a headphone set or signal tracer to follow the signal through the circuit starting at the input. Make sure that you use an isolating capacitor to block d.c. from the headphones or tracer. You should certainly be able to detect the signal at Tr1 collector and Tr2 base. If not you will have trouble in Tr1 stage which, from the foregoing text, should not present much difficulty in locating!

You should also be able to detect signal in this simple way from base to base right up to the output transistors. You may come up against changes in signal level, but the main thing at this stage is merely to trace signal to find out where it stops. When you reach a stage through which signal fails to pass, then more detailed d.c. tests and possible transistor changing should soon expose the defunct component or circuit connection. In the latter respect remember that printed circuit conductors *can* fracture and that the break may be so fine as to be invisible by the unaided eye (use a watchmakers glass or an ohmmeter to establish continuity of a suspect printed circuit conductor or, indeed, soldered connection).

## Signal tracer

A signal tracer consists essentially of an audio amplifier and speaker or socket for connecting an external speaker or headphone set. It is desirable for the input 'probe' to have a high impedance so as not to load the test circuit unduly. Such an instrument could readily be made by the practising service technician or experimenter, representing a relatively small challenge in these days of the i.c. and f.e.t. The i.c. could be engineered to provide the gain, possibly adjustable by an input attenuator or switchable feedback loop, and the f.e.t. to yield a very high input impedance.

The instrument could also be designed to incorporate a meter type of signal sensing in addition to the headphones or speaker, thereby allowing the rapid detection of h.f. signals above the audio spectrum, such as from the local oscillator of a radio receiver or from the bias oscillator of a tape recorder.

An oscilloscope, of course, is another instrument which readily reveals the presence of audio and h.f. signals up to the frequency limit of its Y input amplifier or amplifiers (see, for example, my companion book *Servicing With The Oscilloscope*, Newnes Technical Books).

## Hum test

It is often possible to determine whether or not an audio amplifier is responsive very simply without instruments merely by touching an input point with the tip of a screwdriver upon which a finger is resting. Assuming that the speaker is connected, the amplifier switched on and the volume control advanced a little (that is, when an input point before the volume control is selected), a buzz or hum is produced from the speaker when the amplifier is working reasonably well.

The injected hum input signal is picked up by the person's body making the test from nearby mains power supply circuits. The induced hum voltage is quite small, but due to the high gain of the amplifier when it is working correctly the fundamental frequency and harmonics of the 50 Hz mains supply (60 Hz in some countries, notably American) are amplified sufficiently to result in some sort of output from the speaker.

If this test is done on battery powered equipment at a site well removed from mains supply cables there will not be a hum response (simply because under these conditions there would be no hum signal entering the amplifier!). Nevertheless, it is still often possible to detect a 'click' from the speaker as the screwdriver tip is scraped on an input point.

A word of warning: do not attempt the hum test without thought on a powerful hi-fi amplifier, for the amount of gain and audio power potential at the output could well turn the cone of the woofer inside out, while the 'click' test might produce sufficient transient urge to destroy the tweeter. Wherever possible always select an input point before the volume control and keep the control turned well down. With powerful amplifiers you should really make a rule never to connect or disconnect inputs with the amplifier switched on.

The amplifier should also always be switched off when changing components, and it is best to use a large wattage resistive load across the output instead of the speaker! The resistance of the load should simulate the impedance of the speaker – a value of 8 ohms usually suffices.

After subjecting an amplifier circuit to the d.c. tests described in Chapter 2 and to the signal tests described above, the fault should not remain hidden for long.

## Poor frequency response

Faults other than complete failure might need a little more patience to seek out. Poor frequency response is one such fault condition. Let us suppose that the amplifier in Fig. 4.1 is badly lacking in low bass. This would obviously imply that while the higher frequencies are getting through unhampered the lower frequencies are attenuated. The first places to look would be capacitive couplings. A capacitor blocks d.c. but passes audio (e.g., a.c.). The level of audio signal passed from one stage or circuit to another through a capacitive coupling is related to the reactance ($X_c$) of the capacitor. This is given by $10^6/2\pi fC$, where $X_c$ is in ohms, $f$ the signal frequency in Hz and $C$ the capacitance value in microfarads (µF). This basic formula shows that $X_c$ decreases with increase in frequency.

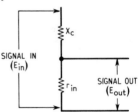

Fig. 4.2 Showing how the input voltage of an interstage coupling is reduced by the potential divider effect of $X_c$, the capacitive reactance of the coupling capacitor, and $r_{in}$, the stage input impedance. The output falls with decrease in frequency since $X_c$ increases as the frequency is reduced

Look at Fig. 4.2, where $X_c$ is denoted by a resistance symbol which, obviously, must appear in series with the impedance of the circuit to which the capacitor is coupled, labelled $r_{in}$. This gives us a simple potential divider, where the output signal ($E_{out}$) across $r_{in}$ is a function of the ratio of $X_c/r_{in}$. It is thus inevitable that the coupled output signal falls with decreasing frequency simply because $X_c$ rises as the frequency falls. Let us assume that $X_c$ is very low compared with $r_{in}$ (e.g., high signal frequency), then $E_{out}$ would be very close to $E_{in}$.

Now consider a low frequency where $X_c$ rises, say, to the value of $r_{in}$. If $X_c$ were purely resistive instead of reactive under this condition

$E_{out}$ would be half $E_{in}$. At that particular frequency the response would then be 6 dB down. This is because a voltage ratio of 0.5:1 corresponds to −6 dB. A voltage ratio of 2:1 corresponds to +6 dB (usually written merely as 6 dB). However, because the current through a capacitor is 90° out of phase with the voltage across it, the net phase shift of a simple RC circuit is −45° at the frequency where $X_c = R$. This means that the response is 3 dB down (not 6 dB as in the purely resistive case) when $X_c = r_{in}$ (Fig. 4.2).

To combat early l.f. roll-off, as it is called, the coupling capacitor is of a sufficiently large value so that its $X_c$ does not rise too much at the lowest frequency of interest. This designed-for state of affairs is defected when the coupling capacitor develops a fault which causes a reduction in its value – and this certainly does happen with capacitors, especially certain electrolytic species.

Thus a common cause of weak bass is reduction in the value of a coupling capacitor, which may occur suddenly or progressively, particularly the latter when an electrolytic capacitor is involved.

The coupler between Tr1 and Tr2 in Fig. 4.1 does not need to be of such a high value as the speaker coupler, for example, because of the relatively high input impedance of Tr2 provided by the feedback which, incidentally, also results in a low output impedance so that the output of the power amplifier resembles a constant-voltage source.

The coupler to Tr1 base need not be of a high value either in order to retain a low bass. This is because of the high input impedance of the stage resulting from bootstrapping (see *Radio Circuits Explained*, Newnes Technical Books).

Another cause of poor bass could be low value or open-circuit of a capacitor (often an electrolytic) shunting an emitter resistor (see Fig. 3.6(a) for example). Without such capacitors the stages are subjected to negative current feedback which reduces their gain. A reduction in capacitance, therefore, would result in greater feedback and hence less gain at lower than higher frequencies ($X_c$ again!).

Poor upper-frequency response is investigated later under preamplifiers.

## Distortion

Distortion can also be troublesome in audio amplifiers. The main cause lies in the output stage, where the two output transistors are arranged in class B. This means that their base bias is adjusted almost to collector current cut-off.

If the output transistors are biased so that there is virtually zero quiescent $I_c$, the two half cycles fail to fit together nicely, and the resulting waveform is very much distorted, as shown in Fig. 4.3(b). This gives a disturbing distortion, not unlike speaker rattle, called *crossover distortion*. To avoid this the base bias of the output transistors is adjusted for a standing or quiescent value of $I_c$. Some amplifiers have a preset resistor in the base bias circuit so that quiescent current can be set for the least crossover distortion.

In an earlier chapter we saw that the $I_c$ is somewhat affected by temperature of the collector junction, and that as the junction

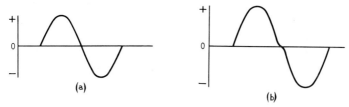

Fig. 4.3. A signal such as (a) appears as at (b) due to crossover distortion, described in the text

temperature increases, so does the effective current passed by the collector electrode. We know that the increase in current is due to collector leakage current.

Clearly, then, if the $I_c$ of a class B output stage is critically adjusted for minimum crossover distortion, a drop in temperature from that when the adjustment was made would be likely to introduce crossover distortion. Unfortunately, this does, in fact, happen. On very cold days or when operating in cold environments, a class B audio amplifier can distort badly until the output transistors have warmed up a bit.

There is a method of overcoming this trouble which takes the form of a temperature-sensitive element in the lower arm of the base potential divider (see Fig. 2.8). The collector junction of a transistor, in fact, being temperature sensitive, can be used for this purpose.

### Bias stabilising diode

The diode or transistor is often connected in a network similar to that shown in Fig. 4.4, and the action is as follows. At low temperatures the diode resistance is high, falling as the temperature increases. Thus, being connected in the lower arm of the base potential divider network of p-n-p transistors, the high 'cold resistance' gives a higher than normal $V_b$, thereby preventing the $I_c$ from dropping to the very low value attributable to crossover distortion. As the ambient temperature rises, the resulting fall in the resistance of the stabilising diode pulls back the negative $V_b$ and thus neutralises the tendency for the rise in $I_c$.

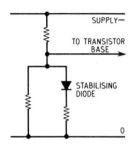

Fig. 4.4. The diode in the bottom arm of the base potential divider has a stabilising action on the base bias, as explained in the text

This sort of stabilisation in particularly useful towards the end of the life of the battery in portable audio equipment (or radio receivers). For the effects of low temperatures on crossover distortion are more pronounced when the supply voltage is below normal, as would be expected, of course. The resistors padding the diode in Fig. 4.4 adjust the overall temperature sensitivity characteristics so that it matches that of the output transistors.

## Preamplifiers and tone controls

A power amplifier needs to be driven from a control preamplifier, and a fairly simple circuit of this kind is given in Fig. 4.5. Preamplification is provided by Tr1/Tr2 and tone controlling by Tr3. The preamplifier also provides for RIAA equalisation which is necessary for use with a magnetic disc pickup. This is achieved by a frequency selective feedback loop, as also the tone control.

Control preamplifiers are often much more complicated than that illustrated, especially super hi-fi ones, and more circuit information on such circuits is given in my *Audio Handbook* (also see Chapter 7). However, the circuit in Fig. 4.5 is useful for getting some good pointers on quick servicing.

The preamplifier part consists of cascoded Tr1/Tr2 both in common-emitter mode. Signal (e.g., a.c.) feedback is provided from Tr2

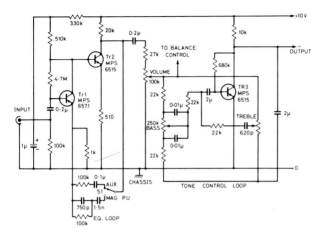

Fig. 4.5. Circuit of control preamplifier of fairly simple design, where Tr1/Tr2 is the preamplifier which can be equalised for RIAA, and Tr3 the tone control stage (see text)

collector to Tr1 emitter through the particular network switched in by S1. With the switch in the 'aux' position the feedback is essentially independent of frequency owing to the relatively high value of the network capacitor whose $X_c$ is swamped at all frequencies of audio interest by the 100k ohm series resistor. With the switch at 'aux', therefore, the gain of the preamplifier is essentially 'flat' over the audio spectrum of interest because the ratio of the feedback loop does not change much with frequency. Gain of the preamplifier on this switch setting is thus mainly established by the 100k resistor and other circuit constants. The smaller the value of the resistor, the greater the feedback over the whole spectrum and the lower the gain.

## Equalisation

In the 'PU' position, however, the feedback loop is modified with change in frequency. This results from the 1.5 nF capacitor and the 750 pF capacitor in parallel with the 100k-ohm resistor, these two forming a time constant of 75 μs. The former capacitor and circuit elements provide the other RIAA equalising time constant requirements (e.g., RIAA *eq* is based on 3180, 318 and 75 μs time constants).

What happens is that as the frequency is increased from a mid-spectrum value, so the impedance of the feedback loop falls which pulls down the gain because there is then an increasing amount of feedback. Conversely, as the frequency is decreased from around mid-spectrum so the impedance of the loop rises which causes the gain of the preamplifier to rise.

These are the requirements for a magnetic pickup whose output increases with recorded frequency. The equalisation ensures that the signal at the output of the preamplifier has a constant frequency characteristic when a magnetic pickup is used. Without such equalisation the preamplifier output from the pickup would rise fairly steadily from a low level at bass to a high level at treble, thereby giving screaming treble and very weak bass.

It is noteworthy that ceramic pickups (of the piezo-electric kind) do not require RIAA equalisation. When such a pickup is playing a disc conforming to the RIAA recording standard (now used universally) the output holds sensibly flat when the pickup is loaded into a high value of resistance. This is because a ceramic pickup responds to the *amplitude* of the recording; unlike a magnetic pickup which responds to the *velocity* of the recording. The RIAA recording equalisation gives a recorded amplitude which is approximately 'flat' with frequency (constant amplitude) and which rises at the approximate rate of 6 dB/octave in velocity.

## Servicing

If you come up against frequency response troubles on magnetic pickups the first place to look is at the equalisation amplifier and time constant components. Sadly, there is no swift way of finding the responsible component, though to make sure that it is not the pickup which is faulty, compare the left and right channels on a stereo amplifier. It would be unlikely for both equalisation channels to develop the same fault simultaneously! Also try a replacement pickup if possible. To get a frequency response plot spot frequencies could be used, and with a constant input voltage you should obtain a characteristic similar to that in Fig. 4.6(b). To obtain a flat response the amplitude characteristics of the input signal should be the inverse of Fig. 4.6(b). It is also possible to obtain RIAA-recorded spot-frequency and swept test discs. By using a pickup to get the output, therefore, an overall response of the pickup and equalisation can be plotted. The actual loading across a pickup can also affect the frequency response. For example, if a magnetic pickup is loaded with a too low value resistor the treble will tend to roll off early.

Audio circuit faults 103

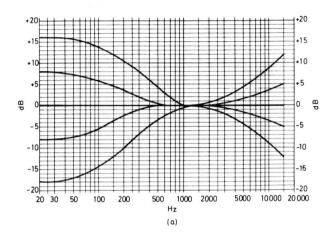

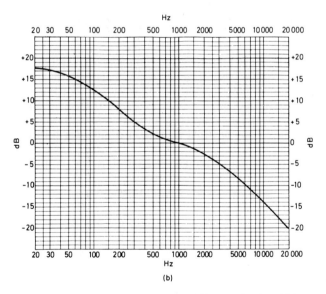

Fig. 4.6. (a) tone control characteristics of Fig. 4.5. circuit and (b) response of the RIAA-equalisation obtained with a constant level input signal

## Tape recorder equalisation

Frequency dependent feedback is also used in the playback amplifier of a tape recorder, for here again the output of the playback head increases with frequency (approximating 6 dB/octave). The equalisation is defined by time constants, as with RIAA record equalisation, which relate to the tape speed and whether the recorder is reel-to-reel or cassette. Pre-emphasis (treble lift) is also applied to the recording signal to help combat high-frequency losses in the head and loss of tape magnetisation as the recorded wavelength is reduced.

## F.M. tuner equalisation

Audio of f.m. radio is given a boost (pre-emphasis) at the treble end at transmission and a cut (de-emphasis) at reception as a means of improving the signal/noise ratio. This sort of frequency tailoring (at the receiver, anyway) is not applied by a feedback loop. A simple $RC$ time constant suffices, the value for UK and Europe being 50 µs and 75 µs in the USA and some other countries.

## Tone control

In Fig. 4.5 you will see that the tone control networks are contained within a negative feedback loop from the collector of Tr3 (via the 2 µF capacitor) back to the base. The networks include capacitors as well as resistors in single-pole $RC$ filter configurations, which means that the maximum rate of change cannot exceed 6 dB/octave. However, the controls allow the degree of boost or cut to be adjusted, and with some circuits it is also possible to switch to different turnover frequencies.

The frequency characteristics of the tone controls of Fig. 4.5 circuit are given in Fig. 4.6(a).

Components changing this values would modify the range of control and possibly introduce asymmetry, making it impossible to achieve the 'flat' response condition with the controls at their centre settings. Like the volume control, the tone controls may eventually become 'noisy', calling for replacement or track cleaning, which is usually not very successful or lasting.

## Noise

Noise in audio amplifiers, assuming that the design is optimised in this respect, often arises from a bad transistor or i.c., particularly in the low-level stages. However, one must not overlook the possibility of a noisy resistor. Resistors can become noisy, especially when passing current, and when such a resistor resides at the front of a high gain preamplifier the noise current is greatly boosted and a background 'hiss' can be heard from the speaker. If this trouble is suspected it can save time by substituting each suspect resistor in turn, having in mind that more than one resistor may be contributing to the total noise.

Note also that the power supply can sometimes reflect noise currents into the early stages. If a noise symptom is persistent, it is a good idea to separate the power supply of the early stages from the main circuit and run the early stages from a new battery (of correct voltage) as a test. If the noise disappears, check not only the supply itself, but also the supply decoupling components.

Electrolytic bypass and coupling capacitors can also become 'noisy'. To discover the stage in which the noise is occurring one needs only to disconnect the coupling capacitors one at a time, reconnecting after each test, starting from that coupling to the base of the last stage. For instance, should the noise be present with a coupling capacitor removed it cannot be occurring in the front stages which are uncoupled. A process of elimination based on this technique will bring to light the defective stage.

When replacing electrolytic capacitors care should be taken to ensure that the polarity is correct. Couplers in p-n-p circuits are connected negative to collector and positive to base. This is because the collector is generally more negative than the base of the following transistor, though there are exceptions to this rule.

Leakage or a bad connection internal to a coupling or output transformer is another cause of noise. Some transformers employed in compact transistor equipment are of sub-miniature dimensions and more susceptible to noise troubles than their larger counterparts.

In direct coupled stages an alteration in value of a circuit element, such as a resistor or even a transistor, can disturb the current distribution in the transistor electrodes. A rise in $I_c$ of the first transistor may result, with a consequent rise in noise. Such trouble, though, would also probably be revealed by a rise in low-level distortion.

Transistors can also develop noise if subjected to an overloading $I_b$ surge.

## Low-level distortion

Low-level distortion, as distinct from the higher level crossover distortion already considered, can arise from alteration in the d.c.

operating conditions of the transistors, due to change in value of a resistor, for instance. An oscilloscope is a useful test-aid for locating distortion. Audio signal from a generator would be applied to the input of the amplifier, and the signal waveform monitored at various outputs along the amplifier chain. Bad distortion would be revealed by the top or bottom half cycle of the waveform being asymmetrical, flattened or rounded excessively.

This test would be facilitated by turning up the input signal to fully drive the amplifier channel. Symmetrical distortion on both half cycles would indicate that the d.c. conditions of the amplifier are well balanced, while distortion below full drive on one or other half cycle would reveal unbalanced d.c. conditions. Whether the base bias would need to be increased or decreased to restore balance would depend upon the mode of the transistor circuit, the type of transistor (i.e., p-n-p or n-p-n) employed and whether clipping was taking place on the positive or negative half cycles of signal.

In circuits employing a base potential divider, whether the bias needs to be raised or lowered can be gleaned by shunting in turn the top and bottom arms of the potential divider by a resistor not less than twice the value of the arm resistor while observing the effect on the monitored waveform. For instance, if the distortion clears or reduces when the bottom arm is so shunted less negative base bias would be required on a p-n-p transistor and less positive base bias on an n-p-n transistor.

One should always have in mind, especially when connecting a test signal to a transistor amplifier, that it is the current wave which is important and, because of the instrinsic non-linearity of the voltage-current characteristic of the emitter junction, a non-sinusoidal input current will result from the direct connection of a sinusoidal voltage. To avoid this effect, the signal voltage should always be applied through a 'swamping' resistor to give a constant current characteristic. It is worth noting also that another cause of distortion in transistor amplifiers is due to the change in current gain of the transistor with collector current. This latter, however, is usually very small and is further minimised by negative feedback.

It is, of course, impossible in the compass of this chapter to look at all the various types of audio equipment. However, much of what has already been said applies from the servicing aspects to all types of circuits.

## Distortion measurement

It is not possible to observe very low level distortion on a sine wave signal displayed by an osciloscope. Meaningful tests require a low distortion sine wave oscillator to supply the input signal and a distortion factor meter to read the total harmonic distortion or, more accurately, the distortion factor, which is the total harmonic distortion plus the noise in the test bandwidth.

The distortion factor meter is a tunable notch filter whose purpose is to remove the fundamental of the test signal, leaving the harmonic and noise components. The voltage of the harmonic and noise components is usually expressed as a percentage of the voltage of the fundamental. To read very low distortion, therefore, the attenuation of the notch must be in the order of 80 dB (corresponding to 0.01 per cent readout) and sharp to avoid attenuating the harmonics to be measured. An audio millivoltmeter is arranged to provide a convenient datum reading of the sine wave signal, the notch is then switched in and tuned, and the sensitivity of the millivoltmeter increased in calibrated steps so that the distortion factor can be read either in dB below the full signal output or direct in percentage. The distortion factor meter includes phasing controls to secure symmetry of the measured signal.

The measurement is enhanced by the use of a dual-trace oscilloscope, one trace displaying the sine wave signal at the output and the other showing the nature of the distortion. The bias presets in the amplifier can then be adjusted for the least distortion over the dynamic range. More information on this method of distortion measurement is given in my book *Audio Equipment Tests* (Newnes Technical Books).

## Another complementary circuit

An early though interesting 2-watt circuit of this kind developed by the Application Laboratory of Thorn-A.E.I. Radio Valves and Tubes Limited is shown in Fig. 4.7 which differs in several ways from the complementary circuit given in Fig. 4.1. Here Tr1 is the input transistor with a variable frequency-selective feedback network giving control of treble. This is in the common-emitter mode and the output at the collector is fed, via the volume control, to the input of a two-stage capacitively coupled amplifier, comprising Tr2 and Tr3, in common-emitter mode.

The collector of Tr3 is directly coupled to the bases of a pair of complementary transistors (Tr4 and Tr5) acting as drivers for the push-pull, class B output pair (Tr6 and Tr7). Direct coupling is also used between the drivers and their corresponding output transistors. The driver Tr5 is effectively in the common-collector mode with the output taken from the emitter while Tr4 is in the common-emitter mode with the output taken from the collector.

Since the driver transistors comprise a p-n-p and an n-p-n type, they conduct on alternate half cycles of signal, thereby driving the output transistors likewise. When Tr7 conducts, signal current flows through the speaker, via R17, and when Tr6 conductors the speaker receives its signal current through R16, the speaker itself being coupled through C9. In effect, then, the driver transistors switch the output transistors on and off alterately.

Stabilising d.c. feedback is applied from the collector of Tr6 to the base of Tr3, via R18 and R19. C10 removes the signal from the network. Frequency-selective and controlled negative feedback, however is introduced between the same points, minus the d.c. (this being isolated by C13). Phase correction is provided by R20 and C11 at the higher audio frequencies. The variable high-pass filter, comprising RV4 and C12, gives a control of bass by allowing for a reduction in feedback towards the lower frequency end of the spectrum, as governed by the setting of the bass control.

Diodes D1 and D2 have developed across them a voltage which biases the driver transistors, this due to the $I_c$ of Tr3 flowing through them in the forward direction. Owing to their characteristics, they make the bias of the driver transistors less dependent upon the $I_c$ of Tr3, as would be the case with ordinary resistors. In effect, the diodes act as a low impedance path for both signal and changes in d.c.

A preset resistor is often found somewhere in the d.c. feedback loop of directly coupled stages to serve as an adjustment for the operating conditions of the output stage. Such a preset is RV3 in the circuit. One way of setting this correctly is to apply an audio signal to the input of the amplifier and adjust its level until clipping occurs in the output stage. The preset is then adjusted until clipping is symmetrical on both half cycles.

Note the high value series resistor, R1, at the input. In practice this would be adjusted to suit the signal source, as already explained.

Considerably less complex audio sections are often found in simple radio receivers, and these are dealt with in Chapter 7.

## Impaired high-frequency response

We have already seen how the low-frequency response can be impaired by a decrease in the value of a coupling capacitor, this increasing the l.f. capacitive reactance ($X_c$) (see page 97), but what about the h.f. response? Designers of super hi-fi amplifiers often arrange for the small-signal response to be flat up to 20 kHz or more. Indeed, I have measured hi-fi amplifiers of this type where the $-3\,\mathrm{dB}$

Audio circuit faults 109

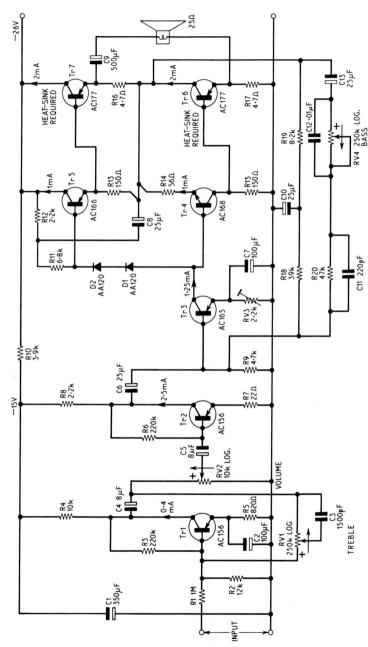

Fig. 4.7. An audio amplifier developed by Thorn-A.E.I. Radio Valves and Tubes Limited using p-n-p and n-p-n transistors in a complementary driver circuit (Tr4 and Tr5) and a single-ended push-pull output stage. This circuit is fully explained in the text

point at 250 kHz. Whether or not this is necessary or even desirable is not a subject for this book! However, for good quality reproduction the response should not start falling significantly before 20 kHz. This is not at all difficult to achieve these days of fast (e.g., high $f_T$) transistors and i.c.s, and the application of negative feedback. Excluding device characteristics, upper-treble roll-off stems from the $X_c$ of shunt capacitance forming a divider with the source impedance. Fig. 4.8(a) shows a skeleton transistor stage, where $R_L$ is the collector load and $C_s$ the following shunt capacitance. From first principles (enough to explain what I mean, anyway), this can be represented by the circuit at (b), where $R_{source}$ is the impedance corresponding to the net effect of the transistor's collector impedance and $R_L$, and $X_{c\ shunt}$ the total capacitive reactance across the output, part of which would be the capacitive reactance of the following stage or circuit.

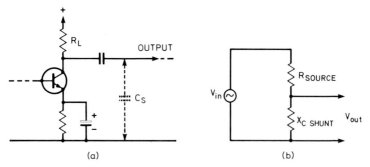

Fig. 4.8. High-frequency roll-off, (a) showing components involved and (b) how these form a potential divider. See text for more details

We already know that $X_c = 10^6/2\pi fC$, where $X_c$ is in ohms, $f$ in Hz and $C$ in µF, which shows that as $f$ increases so $X_c$ reduces. At low frequencies $V_{c\ shunt}$ would normally be very high. Assuming that the actual resistance in parallel with $X_{c\ shunt}$ from the following circuit is much higher than $R_{source}$, then $V_{out}$ would not be much smaller than $V_{in}$.

Now, at higher frequencies the value of $X_{c\ shunt}$ will start to approach the value of $R_{source}$, which means that $V_{out}$ will then start to fall below $V_{in}$. This is from the divider effect of $R_{source}/X_{c\ shunt}$. When $R_{source} = X_{c\ shunt}$ the h.f. response is 3 dB down (also see p.98).

The important point to remember, then, is that if you are feeding audio signal from a high impedance source the parallel or shunt capacitance must be as small as absolutely possible to avoid the upper-frequency response rolling-off too early. This can apply to the self-capacitance of the leads which are used to couple items of audio equipment together. Certain DIN interfaces are prone to this problem owing to the relatively high impedances involved, especially for tape recorder interfaces. It certainly pays the designer to engineer for a low output impedance in radio tuners, preamplifiers, etc. which are likely to be coupled by screened cables to the main amplifier. Screened cables, of course, have a relatively high self-capacitance whose reactance can represent $X_{c\ shunt}$ in Fig. 4.8(b).

## Instability

Before leaving audio amplifiers, a few words should be said about instability and the use of heat sinks. First instability. This can arise due to a common impedance between two coupled circuits. The power supply circuits are vulnerable in this respect, and in battery operated equipment the battery itself developing a high internal resistance towards the end of its life can incite such an unwanted coupling. This is especially so if the supply is not shunted by a high value electrolytic capacitor. The symptom is generally low-frequency oscillation, or 'motor-boating' as is it sometimes called.

Another cause of the symptom is failure of an interstage decoupling capacitor or the capacitor across the power supply line. Failure of an emitter bypass capacitor does not always cause the trouble: this fault usually encourages degenerative feedback (i.e., negative instead of positive).

However, the latter may instigate a phase change in the feedback loop, for instance, the effect of which may then be high-frequency oscillation or oscillation outside the audio spectrum which can be detected best by an oscilloscope.

In high-gain preamplifiers displacement of wiring so that an unwanted coupling exists between the output and input circuits may not only encourage instability but also will almost certainly affect the frequency response towards the higher end of the spectrum. This is less of a problem these days of printed circuit boards.

## Heat sinks

Owing to the dissipation of power at the collector junction, some audio output transistors and all large power transistors are arranged to be in thermal contact with a heat sink. Such a device helps to remove the heat from the junction as quickly as possible, and shows an improvement over the normal method of heat removal which is by a combination of conduction (into the transistor case), convection and radiation.

A heat sink may take the form of the metal chassis upon which the amplifier is built, the heat from the transistor then being conducted into the metal chassis from whence it is lost by radiation and convection. A heat sink allows a transistor to be operated at a greater collector dissipation than would otherwise be possible. Derating curves relating total permissible dissipation to ambient temperature and the type and thermal efficiency of the heat sink are given in data supplied by the transistor manufacturers.

Clip-on cooling fins sometimes act as the heat sink, but whatever their style they serve to keep the transistor as cool as possible under working conditions (like the radiator of a car). We have seen that collector leakage current rises rapidly as the temperature of the junction rises, so that on no account should transistors designed for equipment featuring a heat sink be run at full power in free air. This would almost certainly destroy them since their junction temperature would quickly rise and the rapid build up of $I_c$ would develop more heat and so on.

The metal case of a power transistor is designed for direct clamping to the metal chassis or heat sink. The transistor shown in Fig. 1.11 is of such a design. It should also be noted that the collector of this type of transistor is often terminated to the metal case. Thus, if the design of the equipment does not permit direct connection from the collector to the metal chassis, the transistor must be insulated from the chassis. A mica washer and two bushes satisfy this requirement. Small power transistors are developed by a substantial metal clip acting as the heat sink.

In medium and high power audio stages full protection against thermal runaway cannot be given by the emitter resistor alone, for if the value of this is made too large the power handling capabilities of the stage are impaired. Relatively low value resistors are thus found in the emitter circuits of output stages.

Powerful hi-fi amplifiers are equipped with output transistor protection circuitry. Simplest arrangements use current limiting diodes, while more involved ones use transistors when using abnormally high values of current and/or voltage. Circuits like these delete the audio drive in the event of the power amplifier being coupled to an abnormally low output impedance which would be likely to destroy the output transistors if the drive were maintained. Less elaborate ampilifiers rely on a fuse or fuses which, it is hoped, will blow before the transistors. Should such a fuse fail or the protection persistently operate, check for a short or low value load at the speaker output.

## Video circuits

The video amplifier of a TV set has certain factors in common with an audio preamplifier, though its frequency range needs to be much wider. Monochrome receivers often employ two transistors, the first as driver or buffer and the second as output amplifier which couples to the picture tube.

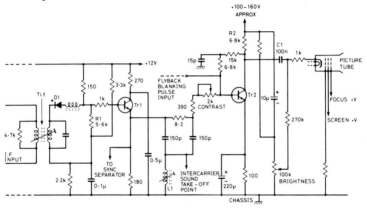

Fig. 4.9. TV video amplifier which is explained in the text. In order to provide sufficient video voltage swing to drive the picture tube, the output transistor Tr2 needs to be powered from a relatively high voltage source

A transistor circuit of this type is given in Fig. 4.9, where diode D1 is the video detector. This receives video-modulated i.f. signal from transformer $T_{i.f.}$ which is the end of a bandpass i.f. channel and which, at the very front, is driven from the TV tuner section. The bandpass and filter characteristics are carefully tailored to suit the video (and intercarrier sound channel) requirements.

The low-level video appearing across detector load R1 is directly coupled to the base of Tr1 which is a common-collector (emitter follower) 'buffer'. Thus the video signal, at the same polarity as at the input, is delivered at low impedance from the emitter. This 'buffers' the detector from the video amplifier stage Tr2.

Video signal is then directly coupled to the base of Tr2, which is a common-emitter amplifier. Video signal swings across collector load resistor R2 are then coupled to the cathode of the picture tube through C1.

D.C. continunity from the detector right through to the picture tube is necessary to minimise phase distortion and to maintain an acceptable l.f. response. Although C1 is used from Tr2 collector for signal coupling to the tube, a d.c. condition is retained through the brightness control and back to the top of the collector load. Different arrangements are found in different receivers, and in some cases the collector of the output transistor is coupled directly to the modulating electrode (cathode or grid) of the tube, but a different brightness control configuration is then required.

In the circuit you will see that the tube grid is at chassis potential through a resistor. The video swing must, of course, operate between grid and cathode, and the signal must be of sufficient amplitude to fully drive the tube whose beam cut-off is around 70 V. Level of signal applied to Tr2 base would probably be in the order of 3 to 4 V black to peak white, which means that the gain of Tr2 would lie in the order of 17 to 23 times (about 25 to 27 dB).

To ensure good picture resolution this sort of gain must prevail over a bandwidth of 4 to 5 MHz. Good phase and transient response are essential to avoid smears and overshoots. In order to achieve the high video swing Tr2 collector load must be fed from a fairly high voltage source, which is sometimes obtained by rectification of the line timebase pulses. Supply voltage requirement of the buffer amplifier is less, of course.

Various artifices are adopted for obtaining an extended upper-frequency response, including the use of high $f_T$ transistors and a small value Tr2 collector load resistor (though there must be a compromise here to avoid abnormally high collector dissipation). For the reason already expounded for audio amplifiers, the source impedance feeding Tr2 base must be low, which is why Tr1 is an emitter follower. If the source impedance were high the h.f. response would fall more quickly owing to the input capacitance of Tr2 and stray shunt capacitances.

As a further aid to an extended h.f. response some circuits include a small choke ('peaking coil') of around 100 µH in series with the collector load. This helps to combat the capacitance at the picture tube modulating electrode (cathode or grid) and the collector capacitance of the heat-sinked output transistor itself. Total shunt capacitance might only be around 13 pF, but for video even this is 'high'.

The choke helps to compensate for this, but too high a value encourages 'ringing' and overshoot symptoms. Other methods of upper frequency compensation might also be found, including passive filters and the use of a smallish value of emitter bypass capacitor which reduces the negative feedback at the higher video frequencies. In Fig. 4.9 feedback is applied by the 15 k and 6.8 k resistors between the collector and base of Tr2. The 15 pF capacitor connected to their junction reduces the feedback at high frequencies so that the amplifier has a gain characteristic which rises with frequency.

The video drive applied to Tr2 base is adjustable by the 2 k potentiometer (wired as a variable resistance), which thus gives a control of contrast. The d.c. between cathode and grid of the picture tube is adjustable by the 100 k potentiometer, which gives a control of brightness. With the control set so that the slider lies at the chassis end of the resistive element and without signal the potential between cathode and grid is around zero. As the control is retarded so the cathode is made more positive than the grid (same as the grid becoming more negative than the cathode); the beam current, and hence brightness falls. Superimposed on this d.c., of course, is the video signal, and it is the negative going picture part of this which pulls the tube away from beam cut-off and illuminates the screen in accordance with the synchronised picture information.

Sync signal is obtained from Tr1 emitter. This is fed to the sync separator stage which deletes the picture part of the signal leaving only the sync pulses. The pulses resulting after suitable differentiation and integration are then sent separately to the line and field timebases for synchronisation.

L1 is the intercarrier trap which is tuned to 6 MHz in the UK. The intercarrier sound signal is picked up from the top of this and fed to the f.m. sound channel. Line and field pulses are applied to a transistor (not shown) and the negative-going pulses yielded by this are connected to Tr2 base. These appear as positive going pulses at the cathode of the picture tube which push the tube deeply into beam current cut-off, thereby suppressing the flyback lines.

One rather critical aspect of the design is the biasing of Tr1. Unless this is set accurately video clipping can occur causing a loss in picture detail. Always check the bias resistors of this stage if you encounter problems of this kind. Some receivers, in fact, include a preset resistor for bias optimisation.

In many receivers nowadays the video detector forms part of an i.c., and i.c.s are used extensively in other parts of the receiver as well.

Video amplifier requirements are less demanding in the camera channel of a closed-circuit TV system or video cassette recorder (VCR) because high video signal swings are not usually necessary. However, good signal/noise ratio is important owing to the relatively small video signal amplitudes delivered by the camera tube. Video at camera output is generally at 1 Vp-p, a value which is also adopted by VCRs at the video output socket. Some cameras include a carrier generator and modulator so that the output may be connected directly to the aerial socket of a TV receiver or VCR.

The servicing of video amplifiers is not all that much different from the servicing of audio amplifiers except that the output is in voltage rather than power (though 75-ohm interfaces are now commonplace) and that the upper frequency response is measured in MHz rather than kHz. It is often possible to check the video amplifier of a TV receiver in situ by applying an audio signal to the input and observing the effect on the screen (particular care must be taken when working with TV receivers owing to the fact that the chassis is commonly in communication with one side of the mains supply; unless suitable mains isolation is employed the result could be lethal both for the connected instruments and, more important, the operator!). A 400 Hz input will produce a bar pattern, full contrast of which should be obtained with an input of about 0.5 V. This will not give much more information than that the amplifier is working and that it is not distorting when an oscilloscope is used to monitor the output wave form. More detailed tests are called for to determine things like frequency response, phase distortion and so forth.

### Colour receivers

Colour television receivers now employ transistor and/or i.c. video stages, including the Y amplifiers, colour-difference amplifiers and, when RGB drive is adopted, primary colour amplifiers. Transistors have been developed to provide sufficiently large video signal swings for the picture tube. Fault finding in the Y amplifier stages is approximately the same as described for monochrome video amplifiers. Indeed, the Y channel of a colour receiver performs essentially the same function as the video channel in a monochrome receiver. The Y channel, however, is generally more elaborate since it incorporates a delay line and chroma notch filter, which is sometimes switched out automatically by a diode or transistor on monochrome transmissions. The switching-in function on colour is generally achieved by the so-called 'ripple' signal emanating from the phase detector as the result of the swinging bursts.

When colour-difference tube drive is employed the Y signal goes, via drive presets, to the tube cathodes, while the colour-difference signals then go separately to the red, green, and blue grids of the tube. When RGB drive is used, however, the colour-difference signals are separately matrixed with the Y signal to yield primary colour signals such that $Y + (R - Y) = R$, $Y + (G - Y) = G$ and $Y + (B - Y) = B$. The matrixing is performed in resistive networks associated with transistor stages. With colour-difference drive, of course, the matrixing is performed by the tube guns, such that $(R - Y) - (-Y) = R$, etc., so that each beam is modulated only with its corresponding primary colour signal.

With primary colour drive the RGB signals are applied separately to the associated tube cathodes or, sometimes, grids, with the electrodes not being driven 'locked' to a d.c. datum, probably via the brightness control and beam limiting circuits.

It is necessary to 'clamp' the driven electrodes to a d.c. level related to the video signal, and this clamping action is performed usually by diodes, one or one set for each drive circuit, which are switched by pulses derived from the line timebase.

It is obviously impossible to detail much about colour television receiver servicing here, and readers interested in this subject might find my book *Colour Television Servicing* (Newnes Technical Books) of some help.

With colour television servicing, as well as monochrome, quite a lot about the performance of the video stages can be gleaned from the nature of the tube display. However, with colour speedy fault-finding is assisted by the three rasters on the screen which together make up the monochrome or colour picture. Complete failure of one primary colour channel gives the picture a specific hue, green failure giving magenta, red failure cyan and blue failure yellow.

Video transistors need to be run from relatively high voltages (this was facilitated in some early transistor receivers by series-connected transistor pairs) and produce substantial video signal swings fully to modulate the tube beams. The devices in this area are thus fairly vulnerable; and this applies also to the clamping devices, which if faulty, encourage hue changes with changes in Y or luminance signal amplitude. I.c.s are used extensively in colour receivers.

This chapter is summarised in Fault Diagnosis Summary Chart 2.

**Fault Diagnosis Summary Chart 2: Audio and Video Amplifiers**

| Condition | Probable cause | Check |
|---|---|---|
| Complete failure | (i) Incorrect d.c. conditions<br>(ii) Lack of input signal<br>(iii) Signal discontinuity<br>(iv) Fuse or protection circuit | (i) D.c. circuits and transistors<br>(ii) Signal input<br>(iii) By hum test and signal tracing. Coupling components. Speaker and its connections<br>(iv) For low loading at speaker terminals |
| Poor frequency response | (i) Defective coupling components<br>(ii) Open-circuit or low value emitter bypass capacitors<br>(iii) Fault in equalising circuits<br>(iv) Excessive shunt capacitance | (i) Coupling capacitors and transformers<br>(ii) Emitter bypass capacitors<br>(iii) Equalising networks, switching and elements<br>(iv) Connecting lead capacitance |
| Distortion | (i) Crossover distortion (output stage)<br>(ii) Low-level distortion (low-level stages) | (i) Quiescent $I_c$ in class B push-pull output stage. Supply voltage. Output transistors. $RC$ network across output transformer. Operation at higher temperature. Negative feedback loops<br>(ii) D.c. conditions. Base bias. Transistors. Coupling capacitors. Bypass capacitors. Equalisation circuits and components |
| Noise | (i) Weak input signal<br>(ii) Faulty transistors<br>(iii) Increase in $I_c$<br>(iv) Faulty equalising | (i) Strength of input signal. Signal input circuits and components<br>(ii) Transistors<br>(iii) D.c. conditions<br>(iv) Equalising |
| Instability | (i) High resistance power source<br>(ii) Faulty decoupling<br>(iii) Faulty emitter bypassing<br>(iv) Fault in negative feedback | (i) Battery<br>(ii) Decoupling electrolytics<br>(iii) Emitter electrolytics<br>(iv) Negative feedback components |

# 5 Fault-finding in R.F. circuits

By the use of tuned filters instead of resistive loads, a transistor circuit can be arranged to respond to, and amplify, r.f. signals within the frequency range from 30 kHz (l.f.) to 1 000 MHz (u.h.f.) or more. This range embraces the ordinary medium-frequency broadcast spectrum, the spectrums put over to intermediate-frequencies (i.f.) for a.m. and f.m. receivers, the short-wave spectrum, the v.h.f. spectrum (from 30 MHz to 300 MHz) – containing television Band I, the broadcast v.h.f., f.m. spectrum (Band II) and television Band III – and the u.h.f. spectrum which extends from 300 MHz to 3 000 MHz and which contains television Bands IV and V.

The design of the amplifier, of course, is greatly influenced by the frequency range and nature of the signals that it is to amplify. From the aspect of a theoretical circuit on paper, there may not appear to be a great deal of difference between an amplifier designed, say, for i.f. signals and one designed for v.h.f. or u.h.f. signals. The difference is emphasised, however, when the physical characteristics of the amplifiers are compared. An amplifier designed for i.f. signals will feature conventional dust-cored and capacitively tuned coils while a u.h.f. amplifier may employ a resonant line of a few millimetres in length instead of an ordinary tuned circuit. Differences in component values and the way that the circuit is formed will also be noticed.

Greater liberties can be taken with the physical design of an r.f. amplifier operating, say, below 1 MHz than with amplifiers operating in the v.h.f. and u.h.f. bands. Once across the v.h.f. threshold and into the u.h.f. spectrum, the components and circuit elements are not always what they appear to be. For example, a short length of tinned copper wire connected in the transistor collector circuit may well be more than just a circuit connection. At u.h.f. it could represent a tapped output transformer or tuned filter.

Similarly, although the emitter circuit may, from the ordinary r.f. viewpoint, appear to be untuned, in a u.h.f. amplifier a pi filter could well be formed by the inductance of the emitter lead in conjunction with the emitter capacitance of the transistor and stray circuit capacitances.

### Effect of small inductance of u.h.f.

That this can happen will be appreciated when it is realised that a 100 mm (4 in) length of 1 mm (0.04 in) diameter wire (about 18 s.w.g.) has a self inductance in the order of 0.1 μH. At 1 MHz this has a reactance of well below 1 ohm. But at 500 MHz the reactance is approaching 300 ohms and at 900 MHz it is nearing 600 ohms.

The innocent looking bit of wire, then, which may be seen in v.h.f. and u.h.f. transistor amplifiers, may be doing a great deal more than coupling a pair of circuits electrically. Incorrectly placed, it could result in positive or negative feedback or detuning, depending upon its position in the circuit and the circuit design.

It must also be remembered during the course of fault-finding in v.h.f. and u.h.f. circuits that short lengths of stout copper wire may be purposely introduced into the circuit to tune or stabilise. Thus it is important to avoid removing or bending such curious bits of wire which, on the face of it, may seem to be doing virtually nothing.

In some designs of u.h.f. amplifiers and tuners the signal is coupled from one section to another through a small aperture cut in the screen which separates the sections. Coupling and decoupling (and neutralising) may also be facilitated by small wire loops connected to critical points on inter-stage screens and shields. Similarly, the point on a screen at which a capacitor is soldered may be critical in terms of feeding back a neutralising signal to combat, say, the base/collector capacitance of a transistor, having in mind that the screens possess an inductive value, and that the magnitude and phase of an induced signal change with position on the screen.

One must also remember that u.h.f. signals travel within the first few microns at the surface of a wire or screen. It is for this reason that u.h.f. screens and conductors are plated – silver plating often being used in communication-standard equipment.

It is quite remarkable that a simple, single-stage transistor u.h.f. amplifier can be possessed of so many electro/mechanical problems in terms of design. These can never be revealed on the circuit diagram, but they must always be kept very much in mind when one is testing and servicing such equipment.

## Medium-frequency stages

Before we look at u.h.f. equipment in greater detail a run-through of the r.f. circuits found in a simple radio receiver would be interesting.
 The circuit in Fig. 5.1 shows the self-oscillating frequency changer, i.f. amplifier and a.m. detector stages of a long and medium wave portable receiver. You will see that all the transistors are n-p-n and that the detector is diode D1. Transistor Q1 is the self-oscillating mixer which is arranged in common-emitter mode. Input tuned circuit at the base consists of a ferrite rod aerial (L1) which can be switched by S12 to tune over the long or medium waveband. The collector is loaded, through R3, to a winding on the local oscillator coil L2. Positive feedback required for oscillation is attained by a tapping on a second winding of L2 being fed back, through R4 and C1, to Q1 base by way of the coupling winding on the ferrite rod aerial. Tuning of the oscillator is accomplished very simply merely by the capacitors across the bottom winding of L2.
 The lower frequencies required for long wave are obtained by the addition of C18 which, along with trimmer TR4, shunts the winding when S13 lies in the LW position. In the MW position variable tuning is provided by VC2 (ganged to VC1 which tunes the ferrite rod), with trimming by TR2. In the LW position TR4 is adjusted for a single station, possibly Radio 4 on 200 kHz. The two trimmers TR1/2 are used to obtain the best frequency tracking over the medium waveband, while trimmer TR3 is adjusted for maximum sensitivity on the chosen LW programme.
 The oscillator/mixer transistor Q1 delivers i.f. signal at around 455 kHz from its collector, which is then tuned by the first i.f. transformer T1. Output from this transformer is coupled to the base of Q2, also in common-emitter mode, and retuned by T2 at the collector. Amplified i.f. signal is passed to the base of the second i.f. stage Q3, which is another common-emitter amplifier, the collector of this transistor then delivering i.f. signal to the detector diode D1 from the final i.f. transformer T3.
 Base current for Q1 is provided by R1, for Q2 by R6 and for Q3 by R9. You will see that an emitter resistor is used in each stage and that each one is bypassed by a capacitor. There is no need for the value of these capacitors to be as high as those used in a.f. circuits owing to

the very much higher signal frequencies involved, remembering that the capacitive reactance falls with increasing frequency.

Typical transistor operating voltages are shown on the circuit. Owing to Q1 being an oscillator, the charge due to signal rectification appearing across C1 modifies the base voltage, so an easy way to determine whether the stage is oscillating or not is to meter the voltage across emitter resistor R2 and then apply a short across the lower winding of L2. When this is done there should be a change in emitter voltage, signifying a current change. If this does not happen when such a receiver is being examined for complete failure, it would almost certainly signify that something is amiss with the oscillator.

The signal conditions through the i.f. stages are straightforward, collector matching being facilitated by the taps for the supply voltage on T1 and T2. These ensure that the collectors are not heavily damped, which would otherwise result in low gain and wide tuning. Primary/secondary ratios are chosen to provide the required match at the bases.

Detector load consists of R12 and VR (the 5k-ohm volume control), with C9 bypassing unwanted i.f. signal. The d.c. appearing at the anode of D1 is applied in series with the base voltage of Q2, through R8. This provides a.g.c (automatic gain control), voltage at D1 anode rises, which back-biases Q2 thereby reducing its gain.

## Neutralisation

Neutralising will be found in some circuits to counteract the intrinsic collector/base coupling of the transistors. Without neutralisation some circuits have a tendency to oscillate and produce 'peaky' responses.

Neutralisation merely involves applying feedback which balances out the internal couplings within the transistors. A simple arrangement consists of a capacitor in series with a resistor connected from one side of the tuned collector load back to the base. The correct side of the load needs to be chosen, of course, to avoid aggravating the problem! But there is no difficulty when the tuned collector load is tapped as shown in Fig. 5.1.

'Peaky' responses are also tamed by loading the primary winding of an i.f. transistor, such as provided by R5 in the circuit.

## Tuned-circuit adjustment

In the circuit the i.f. transformers are tuned either exactly to the centre of the passband or to spot frequencies over the passband to

R.f. circuits 123

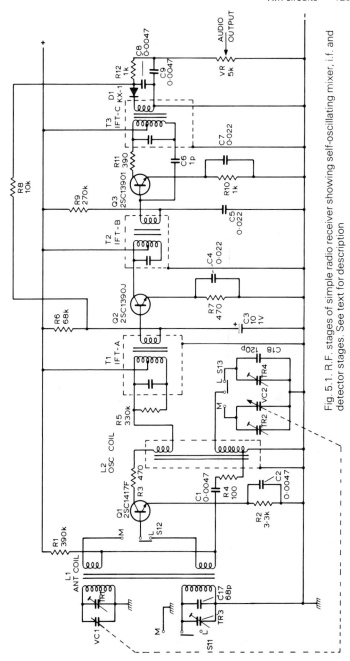

Fig. 5.1. R.F. stages of simple radio receiver showing self-oscillating mixer, i.f. and detector stages. See text for description

yield the required bandpass characteristics. The latter scheme is called *stagger tuning*. When realigning r.f. amplifier stages, therefore, (including i.f. stages, of course), care should first be taken to check whether the tuned circuits should all be adjusted to the same frequency or to slightly different frequencies to provide a wider overall bandwidth. If an amplifier designed for stagger tuning is tuned to one frequency throughout, instability may well result owing to the extra high gain produced. The resulting limited bandwidth may also impair the quality of reproduction where the amplifier is employed in a radio or in a television sound or vision channel. In the latter case, the limited bandwidth would greatly impair the picture definition. Sometimes the bandwidth per tuned circuit is widened by shunting the circuit with a resistor. This, of course, reduces the gain, but is the price that has to be paid for a greater bandwidth.

In the circuit under discussion, each transformer winding is tuned. This is not always the case, for in some circuits tightly coupled transformers are used and then there is only a single tuning core for both windings. At other times the amplifier may feature just a single tuned cicuit, the signal then being extracted or applied via a capacitor.

## Series-connected tuned circuits

The i.f. stages in a.m./f.m. receivers may employ two tuned circuits or transformers connected in series so that the amplifier is responsive to both the a.m. i.f. and the f.m. i.f. without switching. This arrangement is shown in Fig. 5.2., where the a.m. and f.m. if transformers are connected in series in the collector circuit. A similar technique is used for the tuned transformers in the base circuit.

The idea is extended to v.h.f. amplifiers, thereby allowing just one single transistor amplifier to respond to a pre-tuned channel in Band I and a pre-tuned channel in Band III, without the need for switching. The basic circuit of an amplifier of this kind is given in Fig. 5.3. It is possible to enlarge upon this idea to make an amplifier reponsive also to a pretuned u.h.f. television channel.

In all cases, the higher frequency circuits are connected nearest the appropriate transistor electrode. For example, in Fig. 5.2 the f.m. i.f. is 10.7 MHz and this is connected to the collector. The lower frequency a.m. i.f. of 470 kHz is connected after this. Similarly, in Fig. 5.3: here L1 is the Band III coil and L2 the Band I coil.

This method of series connection is posible because of the considerable difference between the frequencies of the tuned circuits.

R.f. circuits   125

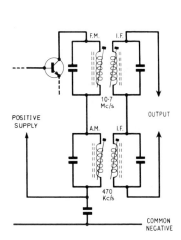 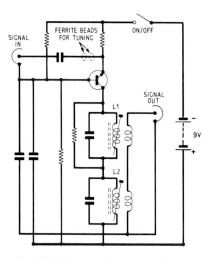

Fig. 5.2. Series-connected tuned transformers as used in a.m./f.m. receivers. Note that the high frequency tuned circuit is connected next to the appropriate transistor electrode

Fig. 5.3. Series-connected tuned circuits are used in some v.h.f. amplifiers for pretuning a channel in Band I and a channel in Band III

When the amplifier is responsive to the lower of the two tuned frequencies, the high-frequency tuned-circuit has a relatively low reactance and does not detract significantly from the efficiency of the lower frequency operation. Conversely, when the amplifier is responsive to the higher of the two tuned frequencies, the capacitance of the lower frequency tuned circuit in series has a low reactance. In practice, it is necessary first to adjust the circuit to the *lower* of the two frequencies and finally to the higher. If the opposite procedure is adopted, the higher frequency adjustment will be destroyed when the lower frequency circuit is adjusted.

## Wideband amplifier

Another type of wideband amplifier is shown in Fig. 5.4. The bandwidth is governed by the wideband transformer in the collector circuit which is formed upon a special ferrite core. This amplifier, in common with that of Fig. 5.3, is arranged in the common-base mode, with the input signal applied to the emitter and taken from the collector. The base is 'earthed' from the signal point of view by the capacitor to the common negative line.

126  Fault-finding in r.f. circuits

Since the emitter signal current is approximately equal to the collector signal current, the power gain ($G_p$) of the amplifier is equal to $R_c/Z_{in}$, where $R_c$ is the load as seen by the collector of the transistor and $Z_{in}$ is the input impedance of the amplifier. In this sort of amplifier the input and output impedance are equal. Thus, $Z_{in}$ equals $Z_{out}$. This means, then, that if the wideband transformer has 1-to-1 impedance ratio the power gain would be unity, because ($R_c = Z_{in} = Z_{out})/Z_{in} = 1$.

In practice, however, the transformer is designed to have a 4-to-1 impedance ratio. This means that the load impedance as seen by the collector is then $4Z_{out}$, thereby providing a power gain of 4 times

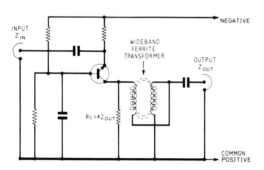

Fig. 5.4. The wideband ferrite transformer in this amplifier circuit serves to transform the output impedance ($z_{out}$) to appear to the collector as four times $Z_{out}$. In this way wideband power gain is possible

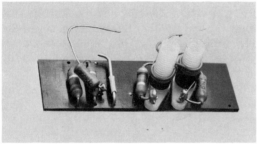

Fig. 5.5. Top view of the printed circuit board insert as was used in the King Telebooster, showing the two series connected tuned circuits

Fig. 5.6. Underview of the printed circuit board of Fig. 5.5. Note the bonding of the transistor case to the mass of common conductor on the circuit board

(6 dB). The bandwidth gain factor can be adjusted by means of the impedance ratio of the transformer, and additional gain can be obtained by cascading two or more stages. With a transistor of high $f_T$, an amplifier of this kind can be designed to give useful gain over the v.h.f. and u.h.f. bands.

The physical makeup of a v.h.f. transistor amplifier is shown in Figs. 5.5 and 5.6. In the former, the top view of the printed board insert is seen, showing clearly the tuned circuits, while in the latter is shown the underside view and the position of the v.h.f./u.h.f. transistor. For reasons of optimum stability, the metal shell of the transistor is held tightly to the mass of common conductor on the board.

### Resonant line u.h.f. amplifier

As already intimated, a somewhat different design technique is demanded for u.h.f. amplifiers. One method is illustrated in Fig. 5.7. The picture shows the top screen removed from the cavity which houses a short resonant line and coupling loop. The ceramic trimmer for tuning the line and the vertically mounted transistor on the right-hand side of the opened cavity are also visible in the picture.

Fig. 5.7. View of a resonant line u.h.f. amplifier with the top screen of the 'cavity' removed to show the resonant line, tuning trimmer, coupling loop and transistor

Here, then, is an example of an amplifier adopting a form of resonant line instead of a conventional coil. Coils are sometimes used in u.h.f. amplifiers, but very small values of inductance and capacitance are required to tune a u.h.f. channel. At u.h.f. strays of capacitance $C$ and inductance $L$ can make tuning to the top channels in Band V somewhat difficult. Most of us know that the frequency of an ordinary tuned circuit is equal to $159/\sqrt{LC}$, where the frequency is in MHz, $L$ is in μH and $C$ in pF. A little calculation based on this formula will reveal the incredibly small values required for $L$ and $C$ to tune, say, to Channel 65 (vision 823.25 MHz). Unless the amplifier is very carefully designed the circuit wiring inductances and the stray circuit capacitances will exceed the computed $L$ and $C$ values.

The resonant line which takes the place of the $LC$ circuit is really a transmission line of specific length. Any transmission line whose *length* is adjusted to correspond to a tuned frequency is effectively the equivalent of a tuned $LC$ circuit. When the line is shorted at one end (as in the amplifier in Fig. 5.7) it is resonant at ¼, ¾, 5/4 (and etc.) wavelength, while an open circuit line is resonant at ½, 3/2, 5/2 (and etc.) wavelength.

The physical length of a tuned transmission line corresponds almost to the tuned wavelength. Thus, at Channel 33 (vision 567.25 MHz) a half-wave transmission line is about 28 cm (about 11 in) in length. For such a line to tune from, say, Channel 22 (470 MHz) to Channel 68 (860 MHz) some complicated arrangement would be needed to vary the physical length of the line from about 32 cm to 17.4 cm. This would be out of the question, of course.

Fortunately, the line can be greatly reduced in physical length while retaining the required electrical resonant length. This is done by cutting off lengths either side and replacing these with capacitance. By this means a half-wave line to embrace the u.h.f. television channels can be reduced to a physical length in the order of 5 cm. Tuning over the u.h.f. channels is then accomplished simply by varying the value of the capacitance at one end of the line. This has the effect of varying the electrical length of the line without the need to vary the physical length.

In practice, one end of a half-wave line is loaded by a tuning capacitor (or trimmer) and the other end by, for instance, the collector capacitance of the transistor. There may also be a trimmer here to augment the collector capacitance or for trimming.

A quarter-wave line works in a similar manner, but here it is necessary to capacitively load one end of the line only, as the other end is short-circuited anyway. The open end is usually connected to the collector of the transistor. The collector capacitance helps with the loading and this is augmented by a fixed capacitance and/or by a trimmer for tuning.

Generally speaking, a quarter-wave line of this kind can be made shorter than a half-wave line and it is more convenient for transistor applications, as the lead from the collector can be arranged to form a *short* section of the inner conductor of the line. Quarter-wave lines are more difficult to employ in a valve circuit, as the conductor to the valve electrode connected to the line may itself represent almost the electrical quarter wavelength. Half-wave lines are thus found in valve circuits.

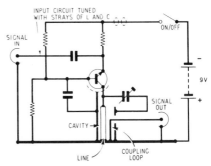

Fig. 5.8. Basic circuit diagram of a tuned-line u.h.f. amplifier showing the connection of the resonant line assembly

The basic circuit of a tuned-line u.h.f. amplifier is given in Fig. 5.8, the cavity and resonant line section with the coupling loop being clearly indicated.

## U.H.F. tuner

Resonant lines are also found in u.h.f. tuners of television receivers, both in the r.f. amplifier and the frequency changer stages. The circuit of such a tuner is given in Fig. 5.9. It will be seen that the collector of the first transistor (r.f. amplifier) is terminated to the first quarter-wave line L1. From this line, signal is coupled into the line L2, via the coupling loops between the two lines. Signal from L2 is then coupled into the emitter circuit of the second transistor (frequency changer), via L4. This rather complex coupling arrangement provides bandpass characteristics to suit the television signals.

It will be seen that both transistors are wired in the common-base mode, with the bases earthed to signal and the signal applied at the emitters and extracted from the collectors.

The second transistor is also an oscillator (i.e., self-oscillating mixer – or frequency changer) which is tuned by resonant line L3. Signal at i.f. is developed across L5 and fed into the i.f. channel of the

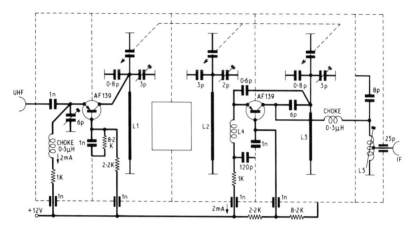

Fig. 5.9. Circuit diagram of a typical quarter-wave u.h.f. transistor tuner. A fourth resonant-line tuned circuit, in the input to the first stage, is generally used in u.h.f. tuners for use in the UK. Most recent TV front-ends use n-p-n transistors, but the principles described remain unchanged

receiver, the signal being derived from the collector circuit. More is said about self-oscillating mixers in Chapter 7 (also see Fig. 5.1). It is worth noting here, however, that the standard British vision i.f. for use with u.h.f. channels is 39.5 MHz.

### A.G.C.

It is possible to vary the gain of a transistor r.f. amplifier by varying the forward emitter current. That is, by varying the base voltage. If the base is made *less* negative (p-n-p transistor), less collector current flows and the effective current gain of the transistor is decreased. This is the '*reverse*' system of gain control.

However, it is also possible to reduce the current gain by *increasing* the emitter junction forward current. That is, by making the base go *more* positive in the case of a n-p-n transistor. This is achieved by the use of a resistor of suitable value in series with the load of the collector. Thus, as the base is made more negative, so the collector current increases, and since the current increase through the series resistor the voltage dropped across it increases and the voltage at the collector falls. This fall in collector volts pulls down the current gain of the transistor stage and thus reduces the signal gain. This technique is called '*forward*' gain control.

To summarise, then, the gain of a transistor varies with both collector voltage and collector current, the former being exploited for forward gain control and the latter for reverse gain control.

The r.f. stages of v.h.f. and u.h.f. tuners now utilise the forward gain control system. Modern v.h.f./u.h.f. transistors are designed with a gain control characteristic that ties to forward control, though not quite in the same way as expounded in the foregoing.

Other gain control mechanisms are now being actively exploited. These are allied to the collector in this way: as the collector current is made to increase by an increase in emitter/base forward current, there occurs an increasing concentration of holes in the collector barrier region and a stronger field in the collector region itself. The effect of this is that the depletion layer widens and the transit time of the current carriers across it is increased, resulting in a decrease in current gain.

The effect is further enhanced by the increased field in the collector region. This results in an increase in collector resistance brought about by the reduced mobility of the current carriers in the stronger field.

From the practical point of view, this technique of gain control, whether manual or automatic, ensures the minimum of circuit detuning, smooth action and a relatively wide control range. It is thus possible to gain-control u.h.f. amplifiers as well as v.h.f. amplifiers which, prior to the techniques described, was virtually impossible due to changing capacitance effects and so forth affecting the tuning and distorting the amplifier response characteristics.

More information on gain control is given at the end of this chapter.

## Power amplifier

This chapter would be incomplete without reference to the r.f. power amplifier, as used in transmitters. These amplifiers are often very similar to their audio counterparts, using a pair of transistors in class B push-pull. Such a circuit is shown in Fig. 5.10.

Here Tr1 is the driver transistor, which usually obtains its drive from a buffer amplifier, this being fed from the oscillator. The driver is coupled to the push-pull pair Tr2 and Tr3 via transformer T1, while the output of the pair is loaded at the collectors by T2. Both the input and output transformers are tuned by C1 and C2 respectvely to the operating frequency.

It will be noticed that bias is not applied to the output pair, since the base circuit is returned direct to the common positive line. This means that under zero drive conditions the current in the collectors is leakage current only. It will be recalled that audio class B output stages demand a little standing bias to eliminate crossover distortion. This, of course, is of no consequence from the r.f. amplification aspect.

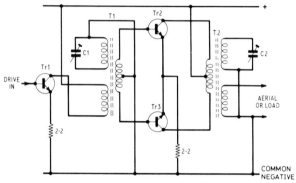

Fig. 5.10. Circuit diagram of an r.f. power amplifier

When a drive signal is applied, the output transistors switch on and off alternatively each half cycle, and signal power is developed across the secondary of T2.

Maximum operating frequency and power output are governed largely by the transistors. As with power audio stages, the output transistors (and sometimes the driver transistors) need to be operated on heat sinks, and it is best if the surface is blackened to help with the cooling.

The driver transistor is also generally operated in class B, and the heat sink is not usually as large as that clamped to the output transistors.

The type of r.f. transformers and their design are governed by the frequency at which it is required for the amplifier to operate. The same applies to the transistors, as with ordinary small-signal amplifiers.

## Fault-finding

Fault-finding in r.f. amplifier circuits should commence by a check of the transistor electrode voltages to establish that the d.c. requirements are fully satisfied, as has already been investigated in past chapters in some detail. Tests of this kind will in many cases reveal the cause of complete failure. However, account must be taken of the effect of strong r.f. signal on the measurements.

## Lack of response – d.c. conditions normal

If the d.c. conditions check normally yet the amplifier fails to respond the fault must be related to the elements in the circuit which are concerned with the signal proper. In power amplifiers leakage or a short across a tuning capacitor is not uncommon. Normally, however, this considerably affects the drive conditions, which in turn affects the biasing of the output transistors. The normal metering facilities will reveal this trouble.

Failure of a capacitor associated with a tuned circuit in a small-signal amplifier can also put the stage completely out of gear without indicating trouble from the d.c. point of view. In the same way, the circuit will be badly put out of adjustment if the overall alignment is disturbed. This can result from an alteration in the setting of the dust-iron tuning cores in the coil and transformer formers. The position of the cores, of course, determines the inductance value of the coil or transformer winding. The inductance increases as more core is screwed into the former, and as a consequence the tuned frequency falls.

Cores are sometimes de-adjusted by unskilled or unknowledgeable enthusiasts, or by vibration due to the equipment being moved from place to place and from forced vibration from an inbuilt loudspeaker (such as in a transistor radio).

If misalignment is suspected as being responsible for the lack of response, the equipment is best checked in the conventional manner with a signal generator and output indicating instrument.

A simple r.f. amplifier can generally be checked for alignment in conjunction with the equipment with which it is employed. An amplifier in front of a radio or television set, for instance, can have injected into it a modulated (or unmodulated, depending upon the type of equipment) signal and with the receiver tuned up and responding, the amplifier tuning can be adjusted for maximum output, as may be indicated from a loudspeaker, cathode-ray tube (i.e., television picture tube) or indicating meter connected to the appropriate position in the receiver.

## Checking gain

The gain of an amplifier can be checked by the use of a signal source and signal indicating instrument. The signal is applied first to the indicating instrument direct and the signal level is adjusted to give a convenient, but low reading on the instrument. The signal is then applied to the input of the amplifier and the output signal is this time fed to the indicating instrument.

If the amplifier has gain, the latter reading, of course, should be somewhat greater than the first reading. The number of times the amplifier output signal level is greater than that of the signal generator gives the 'times' gain of the amplifier. When performing measurements of this nature it is essential that the impedances of the amplifier input and output and the signal generator are equal.

An alternative method of determining amplifier gain is first to set the level of signal from the generator so as to give a convenient reading towards the centre of the scale on the signal indicating instrument with the signal applied to this direct from the generator. Next add the amplifier with the same level of generator signal, and finally add attenuation between the signal generator and the amplifier input until the reading on the signal indicating instrument is the same as when the signal was applied direct. The value of attenuation is then equal to the gain of the amplifier.

This technique can be adopted with any amplifier, and the accuracy is a function of the accuracy of the attenuator and of the accuracy of the impedance matching in and out of the attenuator. A switched attenuator can be useful for this activity, and it is sometimes possible to use the attenuator on the signal generator with quite accurate results.

## Low gain

Low gain, if not caused by mistuning or incorrect circuit alignment, can result from a change in characteristics of a transistor. This can happen due to a short-circuit when voltages at the transistor electrodes are being measured or because of reverse polarity of the battery or power supply. In some cases, a change in characteristics may not be revealed when the d.c. conditions of the stage are checked.

Overloading and damage of a transistor in a v.h.f. amplifier due to transients have been known seriously to impair the effective gain at the top end of the v.h.f. spectrum without in any way affecting the d.c. conditions of the stage. The only satisfactory way of proving this trouble is either by testing the transistor by substitution (that is, by fitting into the circuit a known good specimen) or by checking it in a transistor tester which provides information as to transistor high-frequency performance.

Low gain can also be caused by an emitter bypass capacitor in a common-emitter amplifier going open-circuit or low value.

It is worth noting that apparently low stage gain could result from the action of the automatic gain-control system where the stage is a section of a radio receiver or similar equipment. To avoid misleading results due to a.g.c. action, the a.g.c. line should either be disconnected or short-circuited when gain tests are made. More is said about a.g.c. action in Chapter 7.

Low gain in v.h.f. and u.h.f. amplifiers is invariably caused by some sort of high-frequency trouble in the transistor. However, there are times when a small capacitor becomes lossy with respect to v.h.f. or u.h.f. signals. There is not a great deal that can be done to check capacitors for this kind of trouble, and it generally saves a lot of time simply to check the suspect by substitution, ensuring that it is connected to exactly the same points in the circuit (or on a screen or shield) as the original, that it is of the same type and that the wire ends are of the same length as those on the original component. This applies generally to any replacement made in v.h.f. and u.h.f. equipment. After changing a component in v.h.f. or u.h.f. equipment, retuning is still usually necessary for optimum results.

## Instability in m.f. amplifiers

Instability in medium-frequency amplifiers (such as r.f. and i.f. amplifiers in radio receivers) is nearly always caused by an open-circuit decoupling capacitor. However, if the instability appears to occur only at critical settings of the i.f. transformer or coil tuning, attention should be directed to the components associated with the neutralising or unilateralising networks. A change in value of a capacitor or resistor or an open-circuit in the network could be responsible for the symptom.

If it is found that instability in an i.f. channel, for instance, can be cleared by detuning a tuned circuit, one should not be tempted simply to detune and leave it at that. This is not eliminating the trouble responsible for the effect, and detuning can badly distort the amplifier response characteristics and consequently result in distortion of the signals passed through it. One should make a habit always of getting to the root of the trouble. In the case cited, detuning would simply mask the real trouble.

Another cause of instability is abnormally high gain which may result from trouble in the a.g.c. system, trouble in the base biasing components of the amplifier or from an open-circuit in a feedback or stabilising loop.

It is also worth noting that an increase in the internal capacitance of a transistor can result from overloading, and that this can cause instability.

### Instability in v.h.f. and u.h.f. amplifiers

In v.h.f. and u.h.f. amplifiers suffering from instability, particular attention should be given to the screening between the input and output circuits, for at these very high frequencies a little capacitance between the output and the input is all that is necessary to feed back an in-phase voltage to promote instability or oscillation.

Such amplifiers are also very critical with regard to input and output matching. Design is usually for resistive loads, and if the matching is incorrect the connection of input and output coaxial leads could reflect reactive components across the amplifier loads. This is a frequent cause of instability.

This can be proved by altering the length of the input and output leads. If it is discovered that a certain length of coaxial lead clears the trouble, then one can be pretty sure that a mismatch in the coaxial circuit has a bearing on the fault.

When using such an amplifier near a television set, for example, the greatly increased overall gain between the input of the amplifier and the receiver detector circuits can sometimes be sufficient to incite feedback if the amplifier is placed close to the set or if the input coaxial lead to the amplifier is allowed to pass too close to the detector stages of the set. This lead should always be dressed well clear of the set. The trouble is aggravated if the lead is not properly matched and terminated at the far end. For a similar reason, transistor set-top aerials can result in instability troubles. Some aerials of this nature, however, feature a preset gain control on the amplifier, thereby allowing the gain to be adjusted to a value consistent with stability and good performance.

V.h.f. and u.h.f. amplifiers mounted close to an outside or attic aerial are less prone to feedback troubles because the great distance between the aerial and the receiver avoids the creation of a practical feedback loop.

### Spurious signals

It should be noted, however, that v.h.f. and u.h.f. transistors can generate spurious signals removed from the tuned frequency of the amplifier. Indeed, such transistors have been known to oscillate by themselves when connected in a transistor tester or basic circuit for applying the d.c. potentials. The phenomenon is described in Chapter 2, under the section 'spurious oscillations'.

Under certain, though infrequent, conditions, spurious oscillation can beat with a harmonic of the local oscillator or i.f. signal when a transistor v.h.f. or u.h.f. amplifier is connected in front of a television or radio receiver to boost the aerial signal. While this may not be revealed on the set which is working with the amplifier, the spurious signal so produced can be radiated and it may happen that the frequency of the signal corresponds to the frequency of another radio or television channel used in the area. Interference is thus possible.

Tests have shown that amplifiers situated at the aerial as well as at the set end of the aerial downlead can be equally responsible for this trouble under certain conditions. The curious part about it is that the spurious oscillation occurs only when the set and amplifier are switched on together, and not when amplifier alone is switched on. Certain old-style sets appear to be more prone to this trouble than less vintage models using the 'standard' i.f.

The trouble can often be cured completely by the use of ferrite beads on the power supply wires in the amplifier. These beads are simply threaded over the wires and their effect is to increase the inductance of the leads so that the leads then act as r.f. stoppers or chokes. Another cause of the trouble is excessive inductance on the base of the transistor which is usually in the common-base mode. A change of the capacitor on the base may clear the effect. A low-inductance ceramic or lead-through type is best.

Finally, mention must be made of the possibility of instability arising in v.h.f. and u.h.f. amplifiers and tuners due to disturbance to the wiring and screening or, indeed, to the physical makeup of the equipment generally. A tuner, for instance, operated without its bottom cover, may not only tend towards instability, but its overall efficiency would almost certainly be impaired, as also would its tuning.

Circuit and mechanical liberties just cannot be taken with v.h.f. and u.h.f. equipment, even though such liberties may be possible with medium-frequency equipment without apparent ill effect.

## Overloading and cross modulation

Transistor amplifiers are prone to overloading problems if the input signal is in excess of the signal handling capabilities of the devices used and the circuit design. Overloading in r.f. amplifiers produces cross modulation symptoms. This means that where the amplifier is carrying a number of signals of different frequencies simultaneously, the signals tend to modulate each other, thereby causing spurious intermodulated signals at the output.

A television amplifier, for instance, working under this condition could give sound-on-vision and vision-on-sound symptoms at the receiver. The sound would be revealed by the picture juddering or jumping in sympathy with the modulation, while the vision signal is heard from the speaker as a low-pitched buzz, which tends to change in characteristics with change in picture content. Other intermodulation symptoms are likely to be produced when the input signals differ from those mentioned.

The effect is aggravated by too little forward current in the emitter/base junction, as may result from excessive a.g.c. action or from alteration in value of a resistor in the base potential divider network.

If the trouble arises only when the input signal is excessive, steps should be taken to reduce the level of signal applied to the amplifier by means of an attenuator, designed to match the input impedance of the amplifier.

Transistor v.h.f. and u.h.f. teletuners and amplifiers can also suffer from these problems if the aerial is picking up a large amount of radio or television signal which is not the signal of the required transmission. As the input circuits of transistor amplifiers and some teletuners are wideband, signals in channels adjacent to that tuned may push the amplifier or tuner into overload conditions with consequent intermodulation effects. The solution lies in filtering out the unwanted signals before the wanted signal is applied to the input.

## Gain control

At this juncture, the methods employed for controlling the gain of r.f. amplifiers are worth repeating. There are two basic mthods. One is arranged to *reduce* the base bias current to reduce the gain by reducing the base voltage. This pulls down the collector current and as a consequence tends to impair the signal handling performance of the stage, depending upon the circuit design and the type of transistor employed.

The other method is arranged to *increase* the standing bias current to reduce the gain (forward gain control). Here a decoupled resistor is usually connected in series with the collector circuit, at the 'cold'

## Fault Diagnosis Summary Chart 3: R.f. Amplifiers

| Condition | Probable Cause | Check |
|---|---|---|
| Complete failure | (i) Incorrect d.c. conditions<br>(ii) Faulty tuned circuits<br>(iii) Misalignment<br>(iv) Signal discontinuity | (i) D.c. circuits and transistors<br>(ii) Windings of coils and transformers, and tuning capacitors<br>(iii) Alignment<br>(iv) Signal couplings |
| Low gain | (i) Faulty transistor<br>(ii) Open-circuit bypass capacitor<br>(iii) Misalignment | (i) Transistor and d.c. conditions<br>(ii) Bypass capacitors on emitter and/or base circuit<br>(iii) Alignment |
| Instability | (i) Open-circuit decoupling capacitors<br>(ii) Open-neutralising or unilateralising network<br>(iii) Faulty a.g.c. system<br>(iv) Incorrect or reactive loads (mainly v.h.f. and u.h.f. amplifiers)<br>(v) Unwanted feedback<br>(vi) Spurious oscillation (v.h.f. and u.h.f. equipment)<br>(vii) Misplaced wiring/connections | (i) Decoupling capacitors<br>(ii) Network components<br>(iii) A.G.C. circuit and components<br>(iv) Input and output matching<br>(v) Screening and that input and output circuits are separated<br>(vi) Bypass capacitors and try ferrite beads (see text)<br>(vii) Position and type of replacement parts |
| Overloading | (i) Incorrect d.c. conditions<br>(ii) Input signal too strong | (i) Transistor and d.c. circuits<br>(ii) Signal input level |

end of the tuned load. Thus, the increase in collector current resulting from an increase in base current increases the volts drop across this resistor, which is the same as the collector voltage itself being reduced. In effect, then, the gain is reduced by the reduction in collector voltage brought about by the increase in collector current. This type of control is often used in v.h.f. and (sometimes) u.h.f. amplifiers in teletuners, and in transistor television receiver i.f. strips.

The control voltage for the base is derived either from a manual gain control network (potentiometer) or from a source which provides a.g.c bias, as derived from the level of the signal in the amplifier. More is said about this in Chapter 7.

# 6 Fault-finding in oscillator stages

A transistor oscillator is, in effect, an amplifier with part of its output signal fed back to the input in phase so that after passing through the amplifier the fed-back signal adds to the output signal. Oscillation is sustained by the effective loop gain of the amplifier being equal to or greater than unity.

Transistor oscillators are found in a great diversity of equipment, from ordinary receivers to transmitters, from television timebases to tape recorder oscillators and from audio oscillators to signal generators. The waveforms produced by oscillators are determined by their design requirements. For example, a sine wave output is demanded from an oscillator for application in a radio receiver or tape recorder – the former for the local oscillator and the latter for the bias and erase oscillator – while a sawtooth waveform is generally required from a line or field timebase oscillator or generator of a television receiver and from the timebase of an oscilloscope.

It is not the purpose of this chapter – or, indeed, this book – specifically to study the design of transistor circuits. Here we are concerned as to why a properly designed circuit which was working has ceased to work as it should. Nevertheless, to get the matter resolved we must have some idea of the basic principles of sustained oscillation.

## Tuned oscillators

The most common sinusoidal (sine wave) oscillator is that employing tuned circuits. Such oscillators are employed as local oscillators in

## 142 Fault-finding in oscillator stages

radio sets and as bias and erase oscillators in tape recorders, to quote just two examples.

A simple tuned oscillator circuit is shown in Fig. 6.1. Here signal at the collector across the tuned load L2 C2, is coupled back to the base circuit, via L1, due to the transformer action of T1. The windings are arranged for in-phase feedback.

Resistor R1 applies a little forward current to the emitter-base junction, while C1 earths the 'cold' end of L1 to signal. Since the feedback loop gain of the circuit is greater than unity, oscillations build up quickly when the circuit is powered. Their amplitude is controlled, however, either by the transistor bottoming or by the oscillations producing a bias which cuts off the transistor for a period of each signal cycle. The loop gain of the circuit then becomes equal to unity.

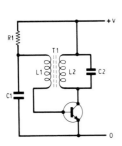

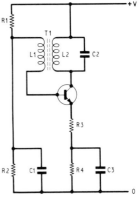

Fig. 6.1. Basic tuned oscillator circuit. Tuning is effected by L2 and C2 and feedback is via L1. Oscillation amplitude is limited by the transistor bottoming. This effect can cause distortion of the output signal

Fig. 6.2. In this circuit the base bias is stabilised by R1 and R2 and the loop gain is controlled by R3. This circuit does not rely so much on the transistor bottoming for amplitude control and the waveform is often purer than that of a bottomed oscillator

In Fig. 6.1 the transistor bottoms due to signal induced into the base circuit from the collector circuit increasing the emitter junction current. The collector current can rise up to a certain value only: thereafter the transistor bottoms and the amplitude of the signal remains constant.

A bottomed oscillator can introduce a relatively high level of distortion and for this reason is not employed for applications where purity of waveform is desirable.

Amplitude control by biasing instead of by bottoming is achieved by the use of a more conventional biasing network and by the use of an emitter resistor or resistors, as shown in Fig. 6.2. Here the mean collector/emitter current is held below the value at which bottoming occurs by the bias and by the limiting effect of the emitter resistors. The unbypassed emitter resistor gives rise to degenerative feedback and is thus used to control the loop gain to provide the purest waveform and to avoid squegging effects.

Capacitor C1 earths the 'cold' end of L1, as before, while R1 and R2 form the base potential divider which functions as the same network used in ordinary transistor amplifiers. C3 is the ordinary emitter bypass capacitor. In some circuits the unbypassed R3 may be omitted, but then the emitter bypass capacitor is chosen to provide the correct amount of loop gain at the operating frequency.

## Collector-emitter feedback

Sometimes the feedback is between the collector and emitter circuits, as in Fig. 6.3. We shall see in the next chapter that this sort of oscillator is popular in ordinary transistor portable radio sets, it being part of a self-oscillating mixer stage (also see Fig. 5.1).

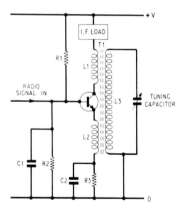

Fig. 6.3. This type of oscillator is found in the self-oscillating mixer section of a transistor radio set. So far as the oscillator is concerned, the base is 'earthed' through C1 and the amplitude of oscillation is to some extent governed by the value of C2. The incoming signal is applied to the base and an i.f. load is incorporated in the collector circuit, as shown

From the oscillator signal aspect, the transistor base is earthed via C1, while the base current is fixed by R1 and R2 as a potential divider. The mean current is limited by R3 and the feedback is controlled by C2. L1 and L2 of T1 are thus coupled, and for tuning L3 is coupled to both of the winding, across which is a variable capacitor.

Transistors of the n-p-n type are often used in modern oscillator circuits, but p-n-p type can be used with equal results, the supply potentials, of course, being reversed in polarity for this type of transistor.

A slightly different collector-emitter feedback arrangement is shown in Fig. 6.4. This is a modified Colpitts circuit where feedback is due to the coupling instigated by C2, between collector and emitter, with the emitter circuit loaded with an unbypassed resistor. Again, from the oscillator aspect the base is earthed, and the transistor can be considered as being in the common-base mode. Such a circuit is found in the oscillator section of some transistor television tuners.

Oscillator signal is coupled out of an oscillator either capacitively from the tuned circuit (or 'tank') or inductively by means of a separate winding. In self-oscillating mixer circuits, the collector is generally loaded with an $LC$ circuit tuned to the difference frequency

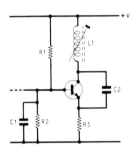
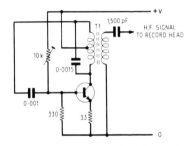

Fig. 6.4. Regenerative feedback between the collector and the emitter is sometimes secured by a capacitance, such as C2 in this circuit

Fig. 6.5. This oscillator circuit is found in transistor tape recorders for generating the record bias signal

between the incoming signal and the local oscillator signal this, of course, being the intermediate frequency.

A practical tuned oscillator as used in a transistor tape recorder is shown in Fig. 6.5. This features variable bias (by the 10k preset resistor in the base circuit) so that the correct signal amplitude consistent with optimum purity can be achieved. Such oscillators work at about 35 kHz and supply an h.f. signal to the record head for bias only. Greater signal power is needed for energising the erase head.

## Push-pull oscillator

Oscillators which provide sufficient power for erasure as well as bias generally use a pair of transistors in a push-pull circuit, as shown in

Fig. 6.6. Positive feedback is arranged between the transistor collector and base circuits via two identical time-constant networks (0.02 μF capacitor and 1.8k resistor).

When the circuit is oscillating the transistors alternately switch on and off over each half cycle of signal, giving a form of class B operation. Feedback coupling is a function of the primary of T1. The secondary of this transformer feeds the signal power into the erase head, this circuit being tuned by the 0.08 μF capacitor for optimum signal power transfer. Signal is also fed from this winding, via a 27 pF capacitor, to the record head as an h.f. bias.

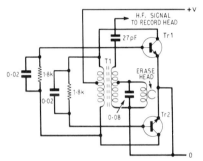

Fig. 6.6. For a signal of sufficient strength to energise the erase head of a tape recorder, a push-pull oscillator of the kind shown here is often employed. This is generally switched to be an oscillator when the recorder is switched from 'playback' to 'record', so that the transistors can be used for the output stage on playback when erasure is not required

It is interesting to note that 1 to 3 watts of signal power is generally demanded by the erase head completely to clear a tape, while only 20 to 50 mW of power is required for biasing. Cassette decks suitable for use with metal particle tape require even more power for good erasure than oxide tapes. They also require a greater h.f. bias.

Lack of complete erase, therefore, should lead to a check of the push-pull oscillator. Low power resulting from mistuning or a defective or worn component is a frequent cause of the trouble. Poor signal/noise ratio after recording on a tape which has previously been erased on the machine should lead to a check of waveform purity. Asymmetricity of the push-pull circuit, due to one transistor being down or alteration in value of characteristics of a component in one side of the circuit, is always a possibility in this respect.

A push-pull oscillator may continue to oscillate even with one of its transistors almost defunct. Under this condition the waveform is usually distorted. However, this condition is often revealed in practice on tape playback because the push-pull transistors and some of the components in the oscillator circuit double in the playback output stage. That is, the design of the tape recorder arranges for the push-pull output stage to serve as a push-pull oscillator when the machine is switched to the record position.

Single-stage oscillators of the tuned circuit variety work by supplying pulses of current, correctly timed, to the tuned oscillatory circuit to sustain oscillation. In some cases the transistors work towards class C, where the transistor is cut off for a substantial period of the cycle.

The oscillator is sometimes biased initially for class A working, but the steady negative voltage (p-n-p transistors) developed at the emitter, due to rectification of the signal when oscillations commence, tends to bias the transistor towards class B working. This gives a stabilising effect on the amplitude of the oscillations, as we have already seen.

## Oscillator tests

From the above, therefore, it will be appreciated that the value of bias voltage on an oscillating transistor will be a little different from that when the same circuit is non-oscillating. This can form the basis of a test to determine whether or not a stage is oscillating. The idea is first to measure the bias voltage across, say the emitter resistor with the circuit switched on normally, and then again with the oscillator tuned circuit heavily damped by shorting out the coil or loading it with a low value resistor. If the stage is oscillating there will be a difference between the two readings. The bias may be either larger or smaller at the emitter, depending upon the type of circuit, when the stage is oscillating. In general, the former condition applies.

Purity of waveform is best ascertained by the use of an oscilloscope. A small amount of the signal at the collector of the transistor should be coupled to the Y terminal of the instrument, and the time-base adjusted to give a display of one or two complete cycles. The coupling should be arranged so that the operating conditions of the oscillator are not unduly disturbed, and the coupling should be as light as possible. Coupling through a capacitor of a few 'puffs' is suitable for most tests of this nature.

The frequency of the oscillation can also be checked on an oscilloscope if the frequency is within range of the capabilities of the instrument. Another way is to pick up the oscillator signal on a radio or television receiver and then obtain a beat note by feeding into the same receiver a signal from a calibrated oscillator or signal generator. The frequency is then read off the instrument at the zero-beat tuning point. More is said about checking the frequency of oscillators on a second receiver in Chapter 7.

Owing to the small number of components in a tuned oscillator stage, there is not a great deal to go wrong. Circuits which are not biased for class A working when they are not oscillating can produce a heavy collector current should something go wrong and stop oscillations or should the load be shorted. For example, the push-pull oscillator shown in Fig. 6.6 would develop a collector current sufficiently large probably to damage the transistors should the secondary of T1 be heavily loaded or short-circuited during operation. Circuits of this nature sometimes incorporate a fuse in the d.c. supply to protect the transistors in such an event.

## Lack of oscillation

Lack of oscillation should lead first to a check of the transistor biasing conditions, then to a check of the transistor and finally to a check of the tuned circuit and associated components. In a circuit such as that in Fig. 6.2 open-circuit or value decrease of the emitter capacitor may reduce the loop gain sufficiently to suppress normal feedback action. Similarly, too great an impedance at the 'cold' end of the base winding of the oscillator coil or transformer may suppress normal oscillation.

If all seems well and yet the circuit persists in its failure to oscillate, a very careful check should be made of the tuned circuit, particularly the coil, for it is not particularly uncommon for a winding to develop short-circuit turns, thereby destroying the action of the tuned circuit.

It is also well worth noting that transistors can develop troubles that impair their gain at high frequencies. Thus, a transistor change would be warranted as an aid in diagnosing a symptom of oscillation cut off at the high-frequency end of the tuning scale, assuming a variable oscillator circuit.

## Squegging

If the collector current pulses are not correctly timed or if their amplitude is excessive, an effect known as *squegging* is likely to occur in the oscillator. The symptom takes the form of a variation in oscillator output which changes at an audio or supersonic (above audio) rate. The situation is aggravated by excessive amounts of feedback.

This can arise due to too great a coupling from the collector back to the base or emitter circuit of the transistor or by the use of too great a value of transistor gain. It is unusual, however, for a normally working circuit suddenly to develop the symptom of squegging. Should this happen after a transistor is replaced, for instance, the loop gain can often be reduced by reducing the value of the capacitor across the emitter resistor.

In radio receivers squegging of the local oscillator has been known to occur after replacement of the self-oscillating mixer transistor, but only at the top end of the m.w. band. This has been proved to be caused by the replacement transistor having a greater gain than the original at higher frequencies. Selecting a transistor which holds stability right up to the top end of the band is one way out of the problem, but probably a better one is to reduce the value of the capacitor across the emitter resistor.

The opposite effect may also occur. That is, oscillator cut off at the top end of the band. This is due to a transistor with less gain than the original. Unfortunately, this condition cannot always be cured simply by increasing the value of the emitter capacitor.

Squegging of the local oscillator in a radio set invariably causes oscillations and beat notes when stations at the top end of the m.w. band are tuned in. If the squegging is at audio, a tone will be superimposed upon the tuned stations.

Blocking can also occur in self-oscillating mixers if the programme signal applied to the base is abnormally high. Such a heavy signal could back-bias the transistor and so cut it off, thereby cutting off oscillations as well.

## Modulation

Sometimes a tuned oscillator is modulated in an item of transistor equipment. A typical example in this respect is, in fact, the self-oscillating mixer. Here the incoming programme signal is applied to the base, while from the oscillator aspect the base is earthed. This is shown in Fig. 6.3.

Audio modulation can also be applied across the base bias resistor to modulate an r.f. oscillator, as in a transmitter, for instance. It is usual in this application, however, to apply such modulation to the base of the r.f. amplifier which itself is driven by an oscillator.

An example in this respect is given in Fig. 5.13 (Chapter 5). Here is shown a driver coupled to a push-pull r.f. amplifier. The practice would be to modulate the driver in this case by applying the modulation signal current in series with the base of Tr1. The driver, of course, is itself an r.f. amplifier operating in class B conditions and thus able to accept modulation.

## Blocking oscillator

Oscillators are also used in pulse-producing equipment. Many such circuits are triggered, a typical illustration being the timebase oscillators of television receivers. Popular in this application is the *blocking oscillator*, a circuit of which is shown in Fig. 6.7.

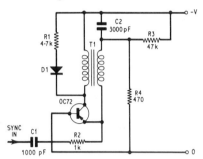

Fig. 6.7. Triggered blocking oscillator circuit. Although a p-n-p transistor is shown, later circuits use an n-p-n device and a positive supply polarity

In this circuit feedback is applied from the collector to the emitter circuit, via the windings on the transformer T1. The transistor is biased by the potential divider R3 R4, while a sync. or triggering signal is applied through C1 and R2.

Now, at the beginning of the first cycle of oscillation the feedback bottoms the transistor. Current continues to flow in the collector winding of the transformer and as a result a constant current flows in the emitter circuit due to a constant e.m.f. being induced into the emitter winding of the transformer. Towards the finish of the cycle of oscillation the emitter current is no longer maintained and the transistor is pulled away from its bottomed condition. Positive feedback then causes the voltage across the windings of the transformer to reverse and thereby switching off the transistor by pushing it hard into collector current cut-off.

At this instant, the charge which has been acquired by C2 is released through R4, and after a period of discharge, determined by the time-constant C2, R4, the pulse-like action repeats.

D1 and R1 across the collector winding of the transformer have the effect of dissipating the energy in the transformer due to the reversal of voltage across the windings.

The circuit shown is triggered by a 1-volt positive-going pulse at the emitter, and it is designed to work at a repetition frequency of 100 kHz.

A triggered blocking oscillator implies that the feedback is insufficient to sustain oscillation and that some triggering device is necessary to instigate the pulse. However, if the loop gain is not less than unity the circuit will be free-running, but it can still be synchronised by the application of pulses, as in a television timebase oscillator. A triggered oscillator is called a 'monostable' circuit, while a free-running oscillator is called 'astable'.

## Astable circuit

Fig. 6.8 shows the circuit of an astable blocking oscillator. Feedback this time is between the emitter and base, via the blocking oscillator transformer T1. The speed control R1 operates by altering the time-constant in the base circuit, thereby allowing the switched-off period of the transistor to be adjusted.

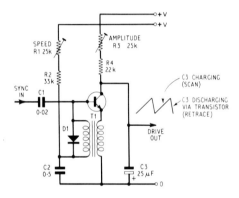

Fig. 6.8. Free-running blocking oscillator circuit

The charging circuit here comprises the amplitude control R3 plus R4 and C3. This gives the time-constant (R3 + R4) × C3. The amount of charge given to C3, and thus the field amplitude, is controlled by the setting of R3.

A sawtooth waveform over the scanning stroke is produced by the increase in voltage across C3 as it charges up; prior to the capacitor reaching anywhere near its full value of charge it is swiftly discharged by the transistor switching on due to the blocking oscillator action. This gives the field retrace and thus completes the sawtooth waveform.

The repetition frequency of the oscillator itself is determined by the time-constant (R1 + R2) × C2. Thus, the repetition frequency is adjustable by the speed control R1, which controls the time constant.

Diode D1 shunts the transformer winding during the retrace period and avoids the induced pulses from damaging the transistor junctions.

There are many variations of the blocking oscillator, but the basic principle of operation of all of them is similar to that described in relation to the circuits in Fig. 6.7 and 6.8. In Fig. 6.8 the frequency is locked by sync. pulses being applied to the base of the transistor. Such pulses instigate the retrace cycle.

In television receivers, the sawtooth waveform is arranged to drive an amplifier stage in such a manner that a linear current is passed through the scanning coils to deflect the spot on the screen of the picture tube either from left to right for the line or from top to bottom for the field. It falls outside the scope of this book, of course, to detail timebase circuits in their entirety. Readers interested in transistor television circuits and their servicing should refer to the author's *Television Servicing Handbook* published by Newnes Technical Books.

## Blocking oscillator faults

Lack of oscillation in a blocking oscillator circuit should first receive the treatment previously detailed for tuned oscillators. Incorrect repetition frequency, or wandering of frequency should lead to examination of the oscillator time-constant components. If these appear to be in order the transformer should come under scrutiny, particularly if the symptom is that of a wandering of frequency probably accompanied by random changes in the amplitude of the waveform. Intermittent troubles in the transformer can sometimes be exposed by pressing the windings of the transformer with the flat side of a screwdriver blade while monitoring the waveform either in terms of scan on a television set or by looking at the waveform properly on an oscilloscope.

Lack of amplitude could be caused either by transistor trouble or by alteration in value of the charging time-constant components. Increase in the value of R4 or decrease in the value of C3 (Fig. 6.8) could cause the symptom. Capacitor trouble, however, may also result in poor field linearity when the oscillator represents the field timebase generator of a television set.

## Multivibrators

The blocking oscillator is sometimes classified as a pulse generator as distinct from a tuned oscillator which creates a sine-wave output.

There are a number of other pulse generators in a variety of different types which contribute to contemporary transistor equipment. It is impossible to explore them all within the compass of this present volume, of course, but one which is frequently encountered and which forms the foundation of many pulse circuits is the *multivibrator* ('multivib' for short), the basic circuit of which is given in Fig. 6.9.

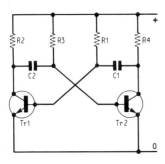

Fig. 6.9. Basic astable multivibrator circuit. See text for full description. In the monostable version, one of the cross couplings is made resistive and base bias is applied to the unstable stage to make it non-conducting until the arrival of a trigger pulse. In the bistable circuit both transistor bases are connected to a separate base bias/trigger input pulse circuit and both cross couplings are resistive. Note that with resistive cross coupling it is, however, usual to connect a capacitor in parallel with the coupling resistor to speed up the action of the circuit

The ouput of this generator is essentially rectangular or square. The term multivib was coined because of the harmonically rich output, having in mind that a square-wave is composed of a large number of harmonically related sine-waves.

The circuit consists of a pair of common emitter amplifiers coupled for regenerative or positive feedback. When the circuit is powered, a slight unbalance in the components and random current disturbances in the circuit push one transistor towards cut-off and the other towards full conduction. This particular state is regenerative due to the cross-coupling. One transistor is switched fully on while the other is switched off, alternating very rapidly, depending upon the circuit time constants, with the reverse condition.

Let us suppose that Tr1 is on (i.e. conducting) and Tr2 is off. This makes the collector end of C2 less negative than the base end. C2 thus charges and, at the same time, the base of Tr2 swings negative bringing it into conduction from its off conduction. Again, the effect is regenerative, so that Tr2 is switched fully on and Tr1 is switched off.

The rate of the switching, which is the repetition frequency of the generator, is governed by C1 discharging through R1, and C2 discharging through R3, and hence by the time-constants of these circuits. There is also a discharge path through the bases of the transistors, so that to some extent the repetition frequency is governed also by leakage currents and, consequently, by temperature.

When the time constants of C1 R1 and C2 R3 are equal, the resulting square wave has equal on/off (i.e., mark/space) times. The mark/space ratio is thus established by the ratio of the two time-constant circuits. Large ratios prevent the circuit from working correctly.

Clearly, there is very little that can go wrong with this type of circuit in its basic form. Lack of oscillation must mean open-circuit of one or more component. The wave shape is influenced somewhat by the repetition frequency, and relatively large rise times could be due to the use of transistors with too low an $f_T$.

## Monostable and bistable circuits

There are two more versions of this sort of circuit. One is a monostable circuit, often called a 'flip-flop' circuit because it has one stable and one unstable state. That is, a triggering pulse is used to 'flip' the circuit into the unstable condition and then subsequently the circuit automatically 'flops' back to the stable state, where it rests until the next triggering pulse occurs.

The other version is the so-called bistable circuit, sometimes given the name Eccles-Jordan circuit after the names of its inventors, when valves were used. This circuit has two stable states, and can be changed from one to the other by triggering pulses. (Note that it is the bistable circuit that is usually called a flip-flop in the U.S.A.)

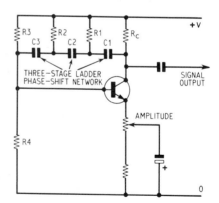

Fig. 6.10. Single-stage phase shift oscillator in which feedback is between the collector and base at the frequency at which the phase is correct for oscillation, as established by the value of the components in the ladder network

These circuits are essentially similar to the multivib but are not fully regenerative. They require triggering pulses to operate them. The astable multivib can also have pulses applied to it to synchronise the repetition frequency, as distinct from operating the circuit.

Another kind of oscillator used in transistor equipment has feedback applied from the output to the input via a phase-shifting network, the reason for which the name 'phase-shift oscillator' is given. A single transistor circuit adopting this principle is shown in Fig. 6.10. Here a three-section ladder network is used between the collector and base circuit to provide the necessary phase shift for oscillation (at one particular frequency the total phase shift introduced by the *RC* networks will be 180°, and at this frequency the circuit will oscillate).

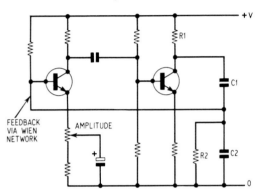

Fig. 6.11. Two-stage Wien-network oscillator. The Wien network or bridge is formed by R1, R2, C1 and C2 between the output of the second transistor and the input of the first

A circuit in which the phase shift is achieved by a Wien network is given in Fig. 6.11. The Wien feedback network comprises R1, R2, C1 and C2, and it is this which provides the phase shift for oscillation. The voltage across C2, R2 has zero phase shift at one frequency only. Positive feedback, resulting in oscillation, occurs at this frequency.

**Fault Diagnosis Summary Chart 4: Oscillators**

| Condition | Probable Cause | Check |
|---|---|---|
| Lack of oscillation | (i) Faulty transistor<br>(ii) Incorrect d.c. conditions<br>(iii) Inadequate loop gain<br>(iv) Blocking<br>(v) Open signal bypass | (i) Transistor<br>(ii) Electrode resistors and power supply<br>(iii) Emitter capacitor, transistor, feedback coupling<br>(iv) Base bias. Signal blocking (mixers)<br>(v) 'Earthing' capacitors or coils of transformer |
| Low output | (i) Low loop gain<br>(ii) Low value charging $C$ or high value $R$ (blocking oscillator) | (i) Feedback coupling, emitter capacitor, supply, voltage, decoupling<br>(ii) Charging capacitor and resistor |
| Squegging | (i) Excessive loop gain | (i) Feedback coupling and emitter capacitor for high value |
| Distortion | (i) Transistor bottoming | (i) Base bias, circuit design (see text) |
| Incorrect frequency | (i) $LC$ circuits out of tune<br>(ii) $RC$ components off value | (i) Tuning<br>(ii) $RC$ components for value |
| Frequency drift | (i) Intermittency in tuned circuit | (i) Tuned circuit components and transformer (blocking oscillator) |

# 7 Fault-finding in transistor radios and hi-fi amplifiers

A transistor radio set comprises r.f. and a.f. amplifiers, a local oscillator, a modulator and a demodulator or detector. The modulator is generally termed the 'mixer'. It receives two signals, the incoming signal and the local oscillator signal. It intermodulates these two signals to produce a difference frequency (usually the oscillator signal *minus* the incoming signal frequency). This is called the intermediate frequency (or i.f.), it being developed across a tuned load or transformer in the collector circuit of the mixer (see Fig. 6.3 in Chapter 6).

In front of the mixer may be an r.f. amplifier, tuned to the frequency of the incoming signal. The majority of the so-called 'popular' transistor sets, however, are arranged for the incoming signal to be applied direct to the mixer, it being only the more sensitive and more expensive models with a pre-mixer amplifier.

The mixer may be self-oscillating or the set may feature a separate stage which operates solely as the oscillator. In the latter case the oscillator signal, along with the incoming signal, is injected into a common-emitter transistor circuit which is designed virtually as a modulator, as mentioned above. Most sets, though, adopt the self-oscillating mixer technique.

The mixer is followed by i.f. amplifier stages which lift the i.f. signal to a suitable level for working the detector stage. This needs a volt or so of i.f. signal. There may be one, two or more i.f. stages, depending on the type, design and sensitivity of the set. Most models feature two stages. The detector rectifies the modulated i.f. signal (the modulation being the actual programme signal, in this case) and produces an audio signal which is fed to an audio amplifier (or direct to a driver stage), then to the driver stage and thence to a pair of transistors in class B push-pull, or alternatively sometimes to a single class A power transistor, which provide the audio power for working the loudspeaker.

Past chapters have described the workings of the various types of amplifiers and oscillators and their faults. This chapter will concentrate on the faults which can actually occur in complete pieces of equipment. For details of how the various sections work and for specific faults and testing procedures, the reader should refer to the appropriate chapter.

By the use of this technique it has been possible in a book of such small volume to embrace a fairly large amount of equipment, having in mind that the majority is composed of little more than amplifiers, oscillators and non-linear devices.

## Non-linear devices

Non-linear devices in the sense meant here are essentially modulators and demodulators (i.e. rectifiers). The mixer stage, for instance, is non-linear. This is why the two signals applied to it are intermodulated. The detector or demodulator has a non-linear characteristic which extracts the audio signal which is modulated on the carrier-wave, whether this be at r.f. or i.f.

A non-linear device (rectifier) is also used in the power supply section to change the mains supply a.c. to direct current for operating the transistors (that is, in equipment which does not rely entirely on batteries for powering).

An amplifier, on the other hand, is designed to be as linear as possible. Non-linearity in an amplifier gives rise to distortion, and this is what the designer does his utmost to avoid. It is impossible to achieve perfect linearity of amplification, but in hi-fi amplifiers the non-linearity is extremely small, since the total harmonic distortion in such equipment is usually below the 1 per cent mark.

## AM/FM receiver

The circuit of an interesting AM/FM battery/mains receiver is given in Fig. 7.1. This is representative of many receivers of this type which do not use i.c.s, though some may not include FM and have a less elaborate audio section. It is a good circuit on which to practise, nevertheless!

158  Fault-finding in transistor radios and hi-fi amplifiers

The various stages are indicated on the circuit so it should not be too difficult to link these to their separate descriptions in the earlier text. In common with almost all radios of this kind, the AM signals are picked up by the ferrite rod aerial. This is tuned only for the MW band because there is no LW AM on this particular model.

## AM section

Looking first at the AM section, then, VT4 is a self-oscillating mixer whose i.f. output is tuned by IFT1–AM. You will see here that the local oscillator tuned circuit is ganged to the ferrite rod tuning and to the r.f. and local oscillator stages of the FM section. A multi-gang mechanical tuning capacitor is commonly utilised, which is coupled to the tuning cursor running along the AM and FM scales.

I.F. signal from IFT1–AM secondary is fed to the base of the common i.f. amplifier stage VT5. An interesting point here is that the signal passes through the secondary of the second FM i.f. transformer IFT2–FM. Note that the FM front-end is unpowered on AM by switch S2A, the same switch powering the AM front-end in the AM mode.

VT5 is a common-emitter amplifier whose collector is loaded to the primaries of the AM and FM i.f. transformers IFT2–AM and IFT3–FM (see page 124). The AM and FM transformers are thus connected in series as explained in Chapter 5. Amplified AM i.f. signal is connected by the transformer secondary to VT6 base, which is another common i.f. stage whose collector is again loaded by series-connected AM and FM i.f. transformers, IFT3–AM and IFT4–FM, each feeding its appropriate detector.

The AM detector is diode W4 whose main load consists of R18, and the a.f. signal developed across this is coupled to the volume control R33 through C49 when switch S2B lies in the AM position. You will see that W4 is biased by the supply rail from the bottom of IFT3–AM secondary. This does several things. It tends to improve the linearity of detection, thereby reducing AM distortion. It also puts a positive bias on VT5 base, and owing to the d.c. output of the detector it also gain-controls VT5 (e.g., a.g.c.). R28/C43/C44 form an i.f. filter network.

The a.f. signal is then coupled through C50 and R34 to the base of the audio input transistor VT7. This is followed by a further amplifier stage VT8, which then drives the complementary output transistors VT10/VT11. VT9 is the output stage bias setting transistor, which keeps the class B quiescent current at the required value (see Chapter 4 for more information on a.f. stages).

Fault-finding in transistor radios and hi-fi amplifiers 159

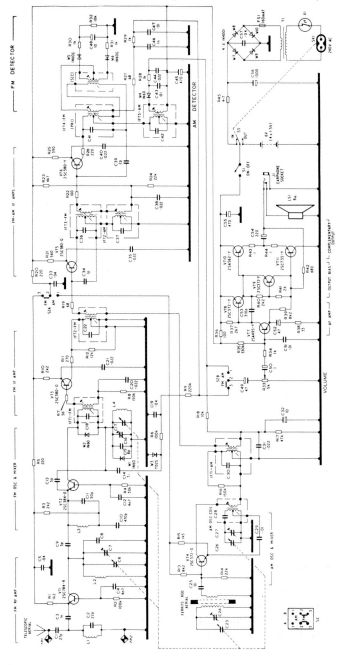

Fig. 7.1. Circuit of AM/FM receiver which can be used on battery or mains supply. This is described in the text along with possible fault conditions

## FM section

In the FM position switch S2A changes the supply from the AM front-end to the FM front-end, while switch S2B changes the audio input to the top of the volume control R33 from the AM detector load to the FM detector load.

R.F. amplifier transistor VT1 represents a good illustration of the common-base mode. Here the FM signal from the telescopic aerial is coupled to the emitter through C1/C3. L1/C2 is an aperiodic (untuned) input filter. Collector of VT1 is tuned by L2/C6/C7, and signal across this is coupled to the emitter of VT2 mixer/oscillator stage, this transistor also being in common-base mode. Note that the bases of VT1/VT2 are effectively 'earthed' to r.f. by C4/C12. You will notice that the chassis or 'earthy' side of the circuit is at supply positive, with supply negative being in communication with the emitters of the transistors, the requirement for the n-p-n transistors since their collectors are eventually returned, d.c.-wise, to the 'earthy' positive line. The bases are also at positive potential through suitable value resistors, which set the forward base currents. Oscillator coil is L4, with tuning by the shunt capacitors, and feedback is by C13. Diodes W1/W2 tend to limit the signal levels due to rectification and hence damping in the event of excessive signal amplitude.

FM i.f. is 10.7 MHz, so in the FM mode only the FM i.f. transformers respond (the AM ones in series have a low impedance to this high i.f.). As with the AM i.f. signal, the FM. i.f. signal is amplified by VT5/VT6 and eventually coupled to FM ratio detector diodes W5/W6, whose load is R32 and stabilising capacitor C48. FM a.f. is developed across C46, from whence it is conveyed to the top of the volume control R33. FM i.f. filtering is by R29/C47, which also provide a degree of FM de-emphasis. Oscillator signal at the hot end of L4 is coupled to diode W3 through C14, and the resulting positive voltage at W3 cathode after filtering by R6/C19 gives a little forward bias to the ratio detector diodes.

Readers requiring more detailed information on the AM and FM detectors are referred to Chapter 5 of *Radio Circuits Explained* (Newnes Technical Books).

## Power supplies

The receiver can be powered either from the mains or a 6 V battery. When the mains plug is inserted switch S3 changes over to disconnect the battery and to bring the mains supply source into circuit. S1 is the ordinary on/off switch.

The mains power supply consists of isolating mains transformer T1 and rectifier bridge W7/8/9/10. Electrolytic C56 is the reservoir capacitor, R45 a current limiter and electrolytic C55 the main smoothing capacitor. Further smoothing and filtering are provided by electrolytic C45. C5/C33 on the negative supply line are essentially r.f. bypass capacitors.

As with most receivers of this type, a facility is provided for the connection of an earphone socket. When the associated jack plug is pushed into socket J1 the speaker is disconnected and the earphone is connected instead to the output of the a.f. amplifier.

X1, incidentally, is a thermal cutout which disconnects the mains supply in the event of excessive mains current flow stemming from a fault condition, such as in the mains transformer, which fails to blow fuse FS1.

## AM alignment

Let us suppose that all the AM tuned circuits are completely out of alignment.

We need two main instruments, a signal generator whose r.f. signal can be modulated with an audio tone and an indicator of audio signal strength. The audio signal indicator is connected in place of the loudspeaker (e.g. the earphone socket J1), in parallel with the loudspeaker or at some other suitable point where audio output signal will cause a response.

A typical arrangement is to use a wattmeter in place of the loudspeaker. This instrument, of course, must be terminated in an impedance which matches that of the disconnected loudspeaker to avoid the output stage from being unloaded. In the circuit under discussion, an output impedance of 8 ohms is required.

An a.c. voltmeter could also be employed. This would be connected in parallel with the speaker so that any increase in audio tone from the speaker would give a corresponding increase in deflection of the a.c. voltmeter. An alternative method is to connect a current meter in series with the battery or power supply to the transistor set. When the set is not delivering a signal to the speaker the current consumption will be nominal. However, as the power fed from the class B output stage to the speaker rises so also will the current consumption. Thus, the loudest output is indicated by the greatest current reading.

The signal generator should be isolated from the transistor set via an 0.1 µF capacitor, and the signal should be fed via the attenuator from the low impedance socket.

## I.F. alignment

The first move is to tune the i.f. transformers all to the same frequency or to stagger tune them according to the design of the set. In the majority of sets it is necessary only to peak all the windings of the AM transformers to the 'standard' British a.m. broadcast i.f. of 470 kHz. This applies also to the set under discussion. FM alignment is a little more involved, but let us stay with the AM section for now.

Thus, the signal is applied to the base of VT4. A fairly strong signal should be applied first so as to get some indication from the speaker and/or output indicator. I.f. transformer IFT3–AM should then be adjusted for maximum output. Transformers IF2–AM and IF1–AM should then be adjusted in that order to further increase the output.

As the circuits are brought into tune it is necessary to reduce the strength of the input signal from the signal generator by adjusting the attenuator controls to avoid overloading. It is best to prevent the output from exceeding about 50 mW when the signal generator is modulated to a depth of 30 per cent by a 400 Hz audio signal.

## Oscillator alignment

When the i.f. tuning is completed the next move is to adjust the r.f. section (this means the aerial circuits and the local oscillator).

The modulated generator signal should, to start this procedure, be loosely connected to the ferrite rod aerial. The idea is to avoid damping the aerial circuit as much as possible while maintaining a reasonable signal coupling. Several turns of wire round the rod is usually sufficient.

With the generator and output meter connected, a signal of about 600 kHz (low-frequency end of the m.w. band) should be injected and then the core in the AM oscillator transformer adjusted for maximum output with the set itself also tuned to 600 kHz or to correspond to the injected signal frequency. Some sets have calibration points marked on the tuning dial or drum.

The generator and receiver should then be tuned to the high-frequency end of the m.w. band (about 1 500 kHz) and this time the m.w. oscillator trimmer, C27, should be adjusted for maximum output. The m.w. oscillator trimming should be finalised by repeating the adjustments at the low- and high-frequency ends of the m.w. waveband.

This completes the oscillator alignment.

## Aerial adjustments

The next job is to align the aerial circuits. This is done again by loosely coupling the generator signal to the set by means of a coil of ten turns of 0.07 mm (22 s.w.g.) insulated copper wire wound on a ferrite rod about the same diameter as the ferrite rod aerial. This ferrite coil is then orientated so that it falls on the axis of the ferrite rod aerial about 1 m (3 ft) from it. Alternatively, the coil can be wound on a wooden frame about 300 mm (1 ft) in diameter. In either case the coil is connected across the output terminals of the signal generator and no direct connection is made from the generator to the set.

A signal of about 600 kHz at the low-frequency end of the m.w. band is applied to the coil and the generator output turned well up. This signal should then be tuned in on the set and the generator output adjusted to give a set audio output of about 50 mW. The pick up sensitivity at this frequency should be optimised by sliding the coils along the ferrite rod aerial for maximum output.

The inductance of the aerial coils is altered as they are slid along the ferrite rod aerial, being maximum at the centre of the rod and reducing at either end. It is usually necessary to soften the wax which secures the coils to the rod by the application of a gentle heat. After establishing the most sensitive position for the coils they should be refixed with the wax.

Maximum sensitivity at the high-frequency end of the m.w. band is obtained by adjusting the aerial trimmer C24 for maximum output with the set tuned to 1 500 kHz and a signal of like frequency applied to the coupling loop.

## Ferrite rod aerials

A few words about the ferrite rod aerial would not be amiss at this juncture. A ferrite rod aerial is a development of the old type frame aerial. The presence of the ferrite rod, however, greatly enhances the signal pick up. This is because the ferrite rod has a high permeability (that is, it has a great attraction to electromagnetic waves) and endows the aerial coils with a high $Q$ (goodness) value.

Electromagnetic radio waves are thus concentrated along the axis of the ferrite rod when the magnetic lines of flux of the signal are in line with the rod. The signal pick up is thus maximum under this condition and minimum when the rod is at right-angles to the magnetic field. The ferrite rod aerial is therefore directional.

A ferrite rod fracture can seriously impair the AM sensitivity. The best plan, if possible, is to replace the broken rod, but it is possible to maintain the performance by repairing a broken rod with a proprietary grade cement that requires no heat treatment. The cement should be applied in accordance with the maker's instructions, and care should be taken to see that the two ends of the rod are set up in complete alignment.

Some rods used have an additional winding connected across a socket for accepting a car-type aerial. The purpose of this winding is to inject to the ferrite rod aerial an external signal when the set is used under screened conditions. For instance, when it is used inside a motor car, the metal body of which screens the rod aerial from the signal.

When the rod aerial is screened in this way, the aerial and set become highly responsive to electrical interference inside the car. The signal/interference ratio is considerably impaired and in addition the set is made even more difficult to use in the car owing to the directivity effect of the ferrite rod aerial. Connecting an external car-type aerial via the loop mentioned tends to neutralise the directly effect of the set's internal aerial to some measure and it also improves the signal/noise ratio.

In some cases, however, the effect is very marginal. This is because a simple external car-type aerial can nowhere near match the efficiency of the set's ferrite rod aerial. Moreover, correct matching between the external aerial and and the set is rarely possible.

**FM alignment**

Alignment of the FM section cannot always usefully be achieved by using a modulated signal and monitoring the level of the a.f. at the output stage. In circuits using a ratio detector, as that in Fig. 7.1, a good idea is to monitor the d.c. voltage across load R32, which will increase as the stages are brought into tune. In the case of a balanced ratio detector the voltmeter can be connected as at A in Fig. 7.2. With a Foster-Seeley type of detector an AM signal generator can, in fact, be used, the output indication then being obtained by detuning the secondary of the detector's discriminator transformer (that is, the transformer feeding the detector diodes). Such detuning (by unscrewing the core of the winding) renders the detector sensitive to AM. The voltmeter used for any of these applications should be pretty sensitive and of high impedance.

A different output indicating technique might be necessary with i.c. FM detectors, depending on the nature of the i.c., but one way is

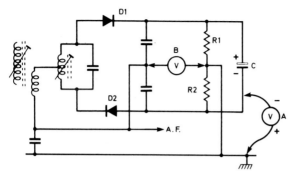

Fig. 7.2. Showing the connection of output meters for FM alignment to a balanced ratio detector

to monitor the output which is available for a signal strength indicator. You will probably find a point somewhere at the end of the FM i.f. channel or around the detector which delivers a d.c. voltage in proportion to the amplitude of the i.f. signal.

## FM i.f. alignment

With capacitor isolation, the 'live' output of the signal generator can be connected to the input of the mixer, such as to the emitter of VT2 in Fig. 7.1. It would be best to mute the local oscillator by putting a short across L4 or the oscillator section of the tuning gang. The signal generator should be tuned to the standard FM i.f. of 10.7 MHz and, with the modulation switched off, the level of the 10.7 MHz signal should be increased until an indication is obtained on the output meter. It is best to reduce the level of the input signal as much as possible consistent with a usable output indication by increasing the sensitivity of the voltmeter (e.g., turning to a lower voltage range). This will avoid errors due to amplitude limiting in the i.f. channel.

The plan then is to adjust the FM i.f. transformers in turn, starting with the last one (but not that feeding the detector), to obtain the maximum output indication, making sure that you turn down the level of the input signal as the stages are brought into tune. Please note that in some cases of severe misalignment it may also be necessary initially to adjust the i.f. transformer feeding the detector to obtain a useful output indication.

Final adjustment to the i.f. transformer feeding the detector should be left to last since this determines the symmetry of response of the carrier on the 'S' characteristic (see Fig. 7.4(e) and (f)). If there is a balanced ratio detector, then a meter can be connected as at B in Fig. 7.2. This will read either in a positive or negative direction depending on the polarity of imbalance. After having adjusted the primary of the transformer for maximum output, the secondary feeding the diodes should then be adjusted for zero output. It will be necessary to alternate between the primary and secondary adjustments to ensure maximum output and maximum symmetry (e.g., zero output from meter B).

With an unbalanced ratio detector, as in Fig. 7.1, a couple of equal value 10 k-ohm resistors can be connected across R32 temporarily to obtain a 'balanced' test point.

Frankly, it is rarely necessary to go to all this trouble for the final balancing can be achieved by amplitude modulating the 10.7 MHz i.f. input signal and monitoring the output from the speaker or other a.f. indicating device. The plan is then to adjust for the least a.f. output, for when correctly balanced an FM detector of this type is the least responsive to AM.

Peaking all the i.f. transformers to 10.7 MHz might not always be the correct way of handling the alignment. This is because the FM sidebands of significance on high quality stereo signal extend up to 240 kHz. These need to be accommodated with the least attenuation and phase non-linearity to maintain the potentially high FM audio quality. The i.f. channel should thus have a bandpass characteristic. With simple mono-only receivers not too much bother is taken about this, and the design might be such that the intrinsic (or deliberately applied by shunt resistance) damping across the tuned circuits provides a sufficiently wide passband. Nevertheless, for the best FM reproduction some form of stagger tuning of the transformers might be called for. It is difficult to optimise in this respect by the simple i.f. alignment described above. Attention should thus be taken of the alignment method suggested by the manufacturer.

### Aligning with oscilloscope and wobbulator

For high quality receivers, especially those designed for stereo reception, a more sophisticated method of alignment is called for, as shown in Fig. 7.3, where the response can be displayed on an oscilloscope. To assist with the alignment a small 'pip' can be produced on the displayed response curve by injecting a marker signal from an accurately calibrated signal generator to the receiver together with the wobbulator signal. The pip occurs on the curve at the frequency to which the marker generator is tuned; it thus becomes a simple matter to determine the overall width of response

Fault-finding in transistor radios and hi-fi amplifiers 167

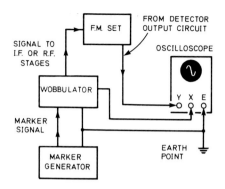

Fig. 7.3. Instrument setup for wobbulator/oscilloscope visual alignment. The Y input to the oscilloscope can be obtained either from the receiver's detector or from a detector probe whose circuit is given in Fig. 7.5

by running the marker pip from one side of the curve to the other and noting the frequency range over which the marker generator has to be adjusted.

Most wobbulators have a terminal available for the injection of a marker signal, but if such a point is not available the marker signal can be loosely coupled to a suitable pick-up point in the receiver itself. Even if an injection point is available on the wobbulator, it is sometimes desirable to connect the marker signal direct to the receiver as a means of securing a pip of large enough size.

Nevertheless, in order to avoid distortion of the displayed response curve it is essential to maintain the smallest possible marker signal consistent with producing a marker pip of workable size. As the marker signal is increased the amplitude of the response curve will be seen to decrease, and as the signal is further increased the marker signal will tend to push the response curve out of shape. During the course of aligning, therefore, it is a good idea to switch off the marker generator occasionally as a means of ascertaining that the shape of the response curve is not in any way influenced by the marker signal.

Whether the Y voltage is picked up from the output of the FM detector or from a diode probe (Fig. 7.5) connected to the output of the final i.f. stage is of little consequence. If it is decided to use the FM detector, however, the secondary of the discriminator transformer should be completely detuned so that the circuit will act as an ordinary a.m. detector, or rectifier, and produce a vertical deflection on the c.r.t. which is proportional to the gain of the i.f. stages.

## Alignment procedures

As alignment procedures recommended by manufacturers will vary according to the circuitry employed in any particular receiver, the following details should be used only as a general guide.

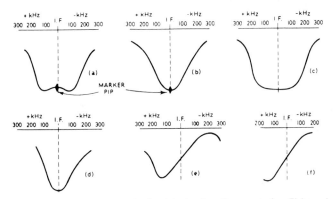

Fig. 7.4. Response traces obtained during the alignment of an FM receiver (see text for details)

The wobbulator signal, at the set's i.f. (10.7 MHz), should be applied to the input of the i.f. stage prior to the discriminator, and the secondary and primary, in that order, of the associated transformer (usually the second i.f. transformer), carefully tuned to procure a response curve of a similar nature to that shown in Fig. 7.4(a). The marker generator should be adjusted accurately to the nominal i.f., and the response curve should be resolved so that the marker pip falls at its centre frequency, as shown.

## Alignment of first i.f. transformer

Alignment of the first i.f. transformer should be carried out with the wobbulator signal applied to the input of the associated stage (normally the FM mixer). Where this point is not readily accessible, the wobbulator signal may be capacitively coupled to the mixer input. The secondary and primary of the appropriate transformer should then be adjusted to produce a response curve similar to that shown in Fig. 7.4(b). It will be observed that this is narrower than the curve at (a) owing to the inclusion of two more tuned circuits. When this curve has been obtained it should be flattened out by readjusting the core in the secondary winding until it looks something like that at Fig. 7.4(c).

With the signal still applied to the mixer input or i.f. test point on the tuner, the core in the primary winding of the discriminator transformer should be adjusted to secure a curve similar to that at Fig. 7.4(d). Finally, the core in the secondary winding, which was initially put out of adjustment, should be adjusted until a symmetrical double curve (Fig. 7.4(e)) is obtained. Optimum symmetry of response occurs only when both diode circuits are in perfect balance, and it is as well to spend a little time over this final adjustment, checking and rechecking for balance and linearity under conditions of both strong and weak signal input.

It should be mentioned that if a diode probe was used for the production of a Y voltage during the alignment of the second and first i.f. transformers, the probe should be removed and the Y voltage picked up from the a.f. take-off point of the dector, usually across the volume control, when adjustment is made to the secondary of the discriminator transformer.

The above adjustments are best made with the wobbulator adjusted to give a sweep of approximately plus and minus 300 kHz. This is, of course, in excess of that required under normal operating conditions, but it does make the response curve of easily workable width.

The curve at Fig. 7.4(f) should be obtained with the wobbulator adjusted for a sweep of plus and minus 100 k, which represents the overall bandwidth required of a correctly aligned FM receiver.

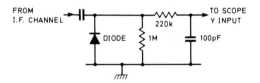

Fig. 7.5. By using a 'detector' circuit like this a signal can be obtained from any part of the FM receiver for application to the Y input of an oscilloscope for visual alignment

It is most important to avoid overloading the receiver or Y amplifier, if used, by applying the smallest possible signal consistent, of course, with sufficient vertical deflection of the trace. The result of overloading will be to produce a curve with a very much flattened top which, unless one is well versed with this mode of alignment, will almost certainly give the impression that the receiver is ideally aligned.

With combined AM/FM receivers, the two alignment processes are totally independent of each other and, unless the makers' instructions stipulate the order in which they should be aligned, it matters little which section first receives attention. Alignment of the AM section follows normal practice.

### FM r.f. alignment

Once the i.f. channel and FM detector have to aligned, the r.f. department is best aligned using an FM signal generator (not forgetting to remove the short from the local oscillator). This should be matched to the aerial input of the receiver and the output monitored on an audio-indicating meter – or the level of sound from the speaker itself could be used as an indicator – but *beware* because amplitude limiting in the receiver i.f. channel can play havoc with the results!

Anyway, first of all the local oscillator should be adjusted to correlate the tuning with the frequency indications on the tuning scale. Using the simple receiver in Fig. 7.1 as an example, about the only thing that you could do here would be to adjust C17 trimmer to get the tuned frequency to correlate somewhere in the middle of the FM band (around 96 MHz, perhaps). Using the smallest level of FM input signal with the generator tuned to 96 MHz, tune the receiver to this signal and note its position on the tuning scale. If it is picked up, say, at 97 MHz you would need to turn the trimmer on a trifle to increase its capacitance and hence reduce the oscillator frequency. Conversely, if the signal is picked up at a setting lower than 96 MHz, you will have to unscrew the capacitor to decrease its capacitance a little.

You should find after this that the tuning will track fairly reasonably over the FM band against the frequency markings on the scale. Some receivers are engineered for the optimising of this frequency tracking. You will sometimes find a core in the oscillator coil as well as the trimmer across the oscillator section of the tuning gang. The plan in this event is first to align at the low end of the band (88 MHz) using the trimmer. Repeated alternations of these two adjustments should result in good frequency correlation over the entire band.

The next move is to bring the aerial and r.f. amplifier stages into tracking alignment. In Fig. 7.1 there is no aerial input tuning, and again there is just one trimmer involved, which is C7 across the r.f. section of the tuning gang C6. I would tune the FM input signal at around 96 MHz, very much reduce the generator signal level (not the modulation but the r.f.) so that the modulation can just about be heard (if you are using a speaker) above the background hiss and then slowly adjust C7 in the appropriate direction to increase the audio and decrease the background noise (e.g., for the best signal/ noise ratio). If you have an audio millivoltmeter sensitive enough to respond to the noise, then turn off the generator modulation and adjust the r.f. trimmer for the least reading. This is by far the best way of tackling the problem, for you will not then be bugged by the possibility of AM limiting. Limiting, incidentally, keeps the audio output fairly constant regardless of changes of level of the FM input signal, so you can see what I am getting at!

A quick way of aligning the r.f. stages of simple receivers of the Fig. 7.1 type if you have no instruments is to tune a fairly strong signal of known frequency (your local one), adjust the oscillator trimmer to bring it to the correct point on the tuning scale, then tune to a very weak signal at the top end of the band (one from a distant transmitter) and finally adjust the r.f. trimmer for the lowest background noise (hiss). If you cannot find a sufficiently weak signal remove the aerial and use instead a short length of wire so that the input signal is at a very low level.

Again, less modest receivers will have provision for r.f. tracking, and in this case use the coil core for optimising the sensitivity at the low end of the band and the trimmer likewise at the top end. Where there are several r.f. stages involved, each one must be separately adjusted as explained, and repeat adjustments will be required to secure the very best results. Always aim for a reduction in background hiss (e.g., the best signal/noise ratio) rather than for maximum audio output.

## Lack of results

We now come to the actual process of fault diagnosis in the simple transistor receiver. A common fault is lack of results. The first move to make in this event is to check the supply voltage and current if necessary. A quick voltmeter test will reveal the state of the battery. If the battery is well up to standard on load (i.e., with the set switched on and taking current), it is possible quickly to establish whether or not the trouble lies in the output stage by listening close to the speaker while switching the set on and off.

If a 'thud' is heard in the speaker when the set is switched on, this is proof that the output transistors are passing collector current and that the speaker itself is in order. If there is no deflection of the speaker cone when the set is switched on, attention should be directed to the speaker proper, its coupling to the output transistors and finally to the transistors and their power feeds.

It is a good idea to acquire the service sheet, manual or, at least, circuit of the receiver under repair and analysis. Either on the circuit or in the manual or service sheet are given the normal voltages at the various electrodes of the transistors. The procedures detailed in Chapter 4 should also be applied to the audio stages.

Whether or not the early audio stages are working properly can be gleaned by touching the base of, say, the first audio transistor with the blade of a screwdriver, with a finger resting on the blade. Life in the audio channel will show up by a click or hum in the speaker as the blade is scraped on an uninsulated conductor in the base circuit.

A distinct fault in the audio section will usually be accompanied by abnormal transistor voltage readings. A defective transistor can also be revealed similarly, as described in Chapter 2.

If the trouble is proved to be in the stages prior to the volume control, a quick run-through of the transistor voltages on the frequency changer and i.f. amplifier stages will often bring the source to light. If the voltages appear to be fairly normal, the local oscillator can be monitored on a second receiver if one happens to be available.

**Local oscillator check**

The idea is to tune the second set to a weak station or carrier and then tune the set under test so that its oscillator corresponds to the frequency of the signal tuned by the second set. The result will be heard as a whistle when the frequency of the two signals are within a few kHz of each other. That is, provided the oscillator of the set under test is working correctly.

It is necessary to secure a reasonable coupling between the oscillator of the set under test and the aerial circuit of the second set. A length of wire connected or coupled to the aerial or ferrite rod aerial of the second set and brought close at the other end to the oscillator section of the receiver under test gives sufficient coupling.

It should be remembered that the local oscillator of the set under test works at the i.f. (usually 470 kHz) above the frequency as tuned on the dial. Thus, if the set is tuned to, say, 1 MHz (300 metres m.w.), the local oscillator should be delivering a signal at 1.470 MHz (a little above 200 metres). A whistle is proof enough that the oscillator is working. Alternative oscillator tests are given in Chapter 6.

The local oscillator of an FM receiver usually operates 10.7 MHz above the tuned frequency, so if the suspect receiver is tuned to

88 MHz you should be able to detect its local oscillator on a second receiver tuned around 98.7 MHz. You will not be able to get a beat whistle in this case, but when you tune to the local oscillator signal the background 'hiss' will fall in level.

If the fault lies in the i.f. stages (or in the r.f. stage if the set has one), testing procedures as detailed in Chapter 5 should be adopted if necessary.

## Symptoms of misalignment

Misalignment may well be reponsible for poor sensitivity and selectivity. The alignment, however, rarely alters by itself, except in a few cases where vibration from the speaker may cause insecure dust-iron tuning slugs to rotate. If visual examination shows that the cores have definitely been altered, possibly by unskilled attention, then a check of the alignment is well worth while.

There are times when a fixed tuning capacitor alters in value, such as those across the windings of the i.f. transformers. Such trouble will show up during an alignment exercise, when it will be found impossible to peak a tuned circuit within range of the core or capacitor adjustment.

Misalignment may also show up by stations failing to tune at their correct points on the dial. Whistles may also be present at various settings of the tuning, and in some cases the set may fail to oscillate at the low-frequency end of the l.w. or m.w. bands, thereby cutting off reception completely and giving blank sections on the dial.

Apparent lack of sensitivity may have its origin in either the i.f. or a.f. stages. To be certain in this respect, it may be necessary to undertake sensitivity checks, as detailed in past chapters.

In many cases, however, lack of i.f. and r.f. sensitivity gives rise to a rather heavy background hiss on tuned stations, even reasonably powerful stations. If should be remembered, though, that very weak signals from distant stations give this trouble on some transistor sets that are without fault.

## Distortion and instability

Lack of a.f. sensitivity is generally accompanied by some kind of distortion. This in itself should lead to a check of the d.c. operating conditions of the audio transistors. Crossover distortion, as explained in Chapter 4, can be troublesome in transistor receivers due to low battery voltage, incorrect biasing of the output pair and to faulty or unbalanced transistors.

Instability can be either in the i.f. or a.f. stages. I.f. instability shows up as whistles as a station is tuned in particularly on AM, while a.f. instability is usually present consistently all over the dial.

I.f. instability is sometimes caused by misalignment or by failure of a neutralising component (see Chapter 5). It can also be caused by unwanted coupling between the output and the input of the i.f. and/or r.f. section. If the instability occurs when the i.f. transformers are peaked during alignment, a check should be made to see whether in fact the transformers should be peaked. Stagger tuning is required by some models. Power supply and feed decoupling capacitors should be examined and tested. Many of these are electrolytic components which tend to alter in value.

## Overloading

Few transistor sets of recent vintage suffer from the effects of overloading due to a powerful signal. However, in the event of an a.g.c. fault, or if the signal is abnormally powerful, blocking may occur when a powerful station is tuned in. This will have the effect of deadening the set completely.

Should this happen when the set is tuned to moderately strong stations, attention should be directed to the a.g.c. circuits.

Some models have an overload diode in addition to the ordinary a.g.c. system. This is connected in shunt with the frequency changer i.f. transformer. It is biased normally so as to be non-conducting, but in the event of a very heavy signal being picked up by the set the bias is changed on the diode by the potential change on the a.g.c. line so that it is biased towards conduction. This, then, damps the i.f. transformer and reduces the signal applied to the i.f. amplifiers from the frequency changer. This technique is adopted because it is not generally a simple matter to apply a.g.c. to the frequency changer transistor, since this is self-oscillating.

The a.g.c. system can be checked for operation by connecting a high resistance voltmeter, set to a low voltage range, between the a.g.c. line and supply positive, noting the voltage when the set is tuned away from a station and seeing whether there is a substantial change in voltage when the set is tuned to a powerful station. The check should be made on the a.g.c. line at the point where it enters the base circuit of the controlled transistor.

Overloading at the front-end of the FM section can cause third-order intermodulation (RFIM for short). Symptoms are 'phantom' or spurious signals resulting from a strong signal appearing at two or more points on the scale, and the modulation of one station appearing on the carrier of a different station, along with its own modulation. This is called cross-modulation. These things can happen even with fault-free receivers when working close to powerful stations. The only cure is to reduce the strength of the aerial signal with an attenuator.

## High quality audio equipment

Let us round off this chapter by looking at some of the circuits featured in the Ferguson (Thorn) '90-series' hi-fi stereo tuner/amplifier/turntable units. There are two models, the 3985 and 3987, the 4-ohm delivery of the former being $2 \times 25$ W and that of the latter $2 \times 40$ W. The power amplifier supply voltages are higher in the more powerful model but the circuitry is similar in both, though for the sake of convenience (and space) not all of it is included in the diagrams.

The amplifier parts are given in Fig. 7.6, the control preamplifier at (a) and the power amplifier at (b). You will now be familiar with some of these circuit sections where they have appeared in isolation in the foregoing chapters.

## Control preamplifier

Looking at Fig. 7.6(a), you will see that the pickup preamplifier consists of VT601/VT602 arranged as a directly coupled pair with equalisation feedback from the collector of the second back to the emitter of the first when S10B lies in the PU position. The other part of S10B at VT601 base is the input selector. S11A selects between FM and auxiliary (aux), while S12A selects FM. With the equalisation of S10B in the other positions the feedback is not frequency dependent since it is then through R610. Main PU equalisation components are R613/R614/R615/C610/C611.

The selected signal eventually reaches the volume control R619 from whence it is passed through the feedback tone controlling stage VT604 by way of emitter-follower VT603. Switch S7A gives a 'loudness' function by virtue of the RC elements involved and the tappings on the volume control. As the control is retarded the output diminishes more at the middle frequencies than at LF and HF so as to produce an effective LF, and to a smaller degree HF, boost for low-level listening.

In Model 3987 the output from VT604 is coupled through C623/C624 to a low-pass (scratch) filter which is a 12 dB/octave, two-pole variety operated by switch S6A. The filtered signal is fed to the base of emitter-follower VT605, which drives a two-pole 12 dB/octave high-pass (rumble) filter when switch S4A is operated.

Power for the control preamplifier section is obtained from the series stabiliser VT702. A similar stabiliser VT703 delivers supply power to other sections of the equipment. The constant-current generator VT701 delivers supply voltage to the tuning unit part of the FM section (not shown). The stepped-down mains supply is delivered by a secondary winding on the mains transformer, T701, and is rectified by diode bridge W705.

Signal for tape recording is obtained from the output of VT602 and fed to the appropriate tags of the DIN sockets – one for normal level and the other for higher level – through the constant-current feed resistor R611. Tape replay signal is coupled direct to the top of the volume control, this signal thereby bypassing the compound preamplifier VT601/VT602, as in this case additional amplification is not required.

It will be appreciated that there are two identical channels for stereo, one for the left signal and the other for the right signal. Only the left-hand channel is shown. Differential gain adjustment between the two channels is provided by the balance control R638.

## Power amplifier

The push-pull output stage can be described as a compound complementary. The compound pair for the positive half-cycles uses VT809/VT811, while that for the negative half-cycles uses VT810/VT812. The speaker is connected direct (no capacitor) to VT811/VT812 collectors on one side and to the zero point of the plus and minus power supply on the other side. When the stage is in balance, therefore, no current flows through the speaker because the offset voltage will be virtually zero.

This condition implies that the voltage at the speaker take-off point is exactly half the supply voltage.

The push-pull output stage is driven from the collector of VT803, which itself is driven from the collector of VT802, and it is to the base of this transistor that the signal from the output of the control preamplifier is applied (via R908/C801/R805).

High-quality audio equipment 177

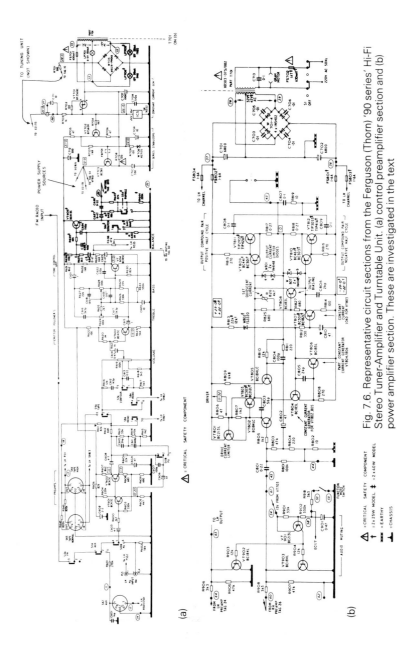

Fig. 7.6. Representative circuit sections from the Ferguson (Thorn) '90 series' Hi-Fi Stereo Tuner-Amplifier and Turntable Unit. (a) control preamplifier section and (b) power amplifier section. These are investigated in the text

Now, with VT802 a long-tailed pair or differential amplifier is formed by the addition of VT805. The base of this transistor, it will be seen, is in d.c. communication with the speaker take-off point at the centre of VT811/VT812 (through R810). This gives 100% d.c. feedback and ensures that the conditions are right for zero offset voltage and hence zero (or very little) d.c. flows through the directly coupled speaker.

A.C. or signal feedback is achieved by the back-coupling from the centre point of the output pair, through R810 and C802, to the base of VT802 by way of the resistive divider formed by R810/R804. Because of the high value of C802 its impedance is very low over the audio range, which means that the a.c. feedback and hence the gain of the amplifier is determined by the ratio of the divider R810/R804.

You will see that R810 is shunted by series combination C804/R811. At high frequencies, which are of little interest to audio reproduction, the impedance of the combination falls thereby turning on more a.c. feedback and hence giving an upper-frequency roll-off to the amplifier.

The differential amplifier VT802/VT805 serves to compare the input signal with the output voltage applied to the speaker. What happens is that an amplified error voltage appears across load R807, which is applied to the base of VT803 driver, the net result giving the centre zero supply voltage condition across the speaker.

The differential amplifier is connected in series with constant-current generator VT804/VT806, while the load for the driver VT803 is also given a constant-current characteristic by VT807.

VT804/VT807 are biased by the current flowing through R813, a state of 'equilibrium' obtaining when VT804 emitter current results in the voltage across R808 corresponding to the switch-on voltage of VT806. This produces a constant current through VT804 collector and a reference potential of around 1.2 V at VT806 collector, this biasing VT807 current generator.

VT807 collector current is established by the value of R814 and by the current flowing through VT804, limited by R808. In the event of a momentarily high speaker current, VT803 driver current is limited by the action of VT801 in conjunction with R806.

Output current is limited by diodes W801 on the positive half-cycles and by diodes W807 on the negative half-cycles. The excess current causes the voltage across R817/R818 to reach the switch-on point of the diodes, either or both thereby switching on and limiting the drive to the compound push-pull transistors.

The higher power model also includes a so-called crowbar protection circuit (not shown). In the event of excessive current flow the charge on a capacitor increases. This causes the firing of a thyristor connected across the power supply rails, which shorts the supply and blows fuse FS702.

The lower power model uses fuses in the negative and positive supplies FS802/FS801 for the right channel and FS804/FS803 for the left channel.

In the event of the fuse on the positive supply line blowing, zener W802 ensures that drive to the constant current generator VT804 is cut off, thereby preventing the complementary output section of the circuit from being driven negatively. The other channel works reciprocally along the same lines.

The biasing of the push-pull output transistors is stabilised by VT808, and is adjustable by the quiescent current preset R826.

## Quiescent current setting

To ensure the least crossover distortion it is essential for the quiescent current to be accurately adjusted, particularly after replacement of the output transistors. This is a basically simple procedure and merely resolves to connecting a d.c. milliammeter in series with either supply line and adjusting R826 for the recommended quiescent current, which for the Ferguson models is 10 to 15 mA.

To avoid breaking a supply line circuit a fuse in the low power model can be removed and the milliammeter connected in place (observing polarity — positive of meter to the positive source). The adjustment should be finalised after allowing a short time for the circuits to stabilise. Of course, there must not be signal drive when measuring the quiescent current.

## Servicing hints

You will see that the supply voltage for the power amplifier is provided by diode bridge W701–4 which is fed from a tapped secondary on the mains transformer T701. The control preamplifier and radio section are powered from separate sources, as we have already seen.

Complete failure, therefore, should first lead to a check of the power supply, making sure that the mains fuse is still intact and also the fuse feeding W705 rectifier bridge in the control preamplifier/radio section.

Should one channel fail, the fact that the other is still operative means that the mains input must be okay. For checking the output stage first disconnect the speaker from the defunct channel and check that both halves of the power supply are active.

If you find a blown fuse, check for a short circuit on the supply line before replacing. You may find that a short developed across one of the output transistors. Also check for open-circuit in the bias transistors.

If you find that there is a d.c. offset at the output check the operation of VT803 driver stage by measuring the voltage across the base/emitter resistor (R806) of the drive current limiter VT801. If this is normal, then you can be fairly sure that the drive and drive current generator stages are working normally. In this event the fault would lie in the differential amplifier.

If the voltage is incorrect, however, a check should be made of the drive current generator by measuring the voltage across R814 emitter resistor. This should be 0.6 V.

If the defunct channel is accompanied with no offset voltage rise check the output transistors. It is fairly safe to reconnect the speaker and make the normal type of signal continuity tests. In the event of high distortion make sure that the quiescent current preset is adjusted correctly. If the fault persists it could be that one of the 0.27 ohm collector resistors has gone open-circuit.

D.C. testing will be required, along with dynamic testing, to bring to light more obscure faults. Remember also to check the working of any voltage stabilising circuits by tracing the voltage from input to output.

**FM i.f. and stereo decoder section**

A fair idea of the circuitry involved in more recent hi-fi equipment can be gleaned from Fig. 7.7. This is the circuit of the FM i.f. and stereo decoder sections of the Ferguson units whose amplifier sections have already been described.

10.7 MHz i.f. signal from the FM front-end is fed to the base of VT201 i.f. amplifier, which is in common-emitter configuration. Amplified signal at the collector is fed to the ceramic filters CF202/CF204, the bandpass-tailored i.f. signal then being applied to the input of the i.f. limiter/FM detector i.c. IC2 at pin 14.

The ceramic filters take the place of the earlier LC tuned transformers, but for correct working require 330-ohm input and output terminations. These are provided by R206/R215. The detector of the i.c. is worked by the tuned circuit L201/C215. The detector of this i.c. operates differently from the ratio detector or similar breed of

High-quality audio equipment 181

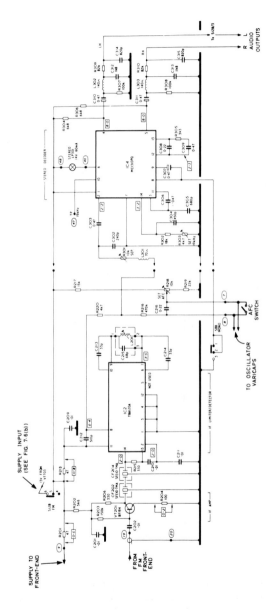

Fig. 7.7. FM stereo radio section (excluding front-end) of the Ferguson '90 series', showing the i.f./limiter/detector i.c. and the stereo decoder i.c., which are considered in the text

two-diode detector. The detector is a so-called 'quadrature' type which means that the phase displacement between two signals is exactly 90°, correcting to a quarter of a cycle in time. This phase shift is achieved by L201/C215. This is merely adjusted for maximum output consistent with the least full deviation distortion (for further information regarding the operation of this and other FM detectors, refer to *Radio Circuits Explained*).

The i.c. also incorporates an automatic frequency correcting (a.f.c.) circuit which can be switched on and off. This ensures that the i.f. carrier lies at the centre of the detector 'S' characteristics, a requirement for the least distortion. The i.c. control signal is fed to a varicap (capacitor diode) across the local oscillator tuning which, with the a.f.c. switch in the position shown, adjusts the tuning for the condition just mentioned. The operating conditions are adjustable by preset R218, the control making it possible to centralise the range of the varicap.

IC2 contains two distinctly separate circuits. One consists of six differential amplifiers, a pair of emitter followers, constant current source and voltage stabiliser. This part acts as the i.f. limiting amplifiers. The other circuit is the quadrature detector already mentioned. The LC circuit is tuned to 10.7 MHz, the standard i.f.

## Stereo separation adjustment

Resulting audio signal (which also carries the stereo multiplex signal) is applied to pin 4 of i.c. IC4, which is the stereo decoder. The block diagram of this i.c. along with associated components is given in Fig. 7.8. The audio/multiplex signal is applied to the i.c. through a low-pass filter L301/R301/C302. This gets rid of spurious signals which might otherwise cause 'birdies' interference. However, phase non-linearity resulting from this filter encourages reduced h.f. stereo separation, and to optimise the separation preset R301 is included. This merely adjusts the phase characteristics of the filter for the best results – e.g., for the least leakage of signal in one channel to the other channel.

The i.c. includes a 76 kHz oscillator in a voltage-controlled oscillator (v.c.o.) arrangement. Divide-down circuits, as can be seen in the i.c. block diagram, eventually provide the 19 kHz stereo subcarrier which was suppressed at transmission but which needs to be reclaimed at reception to decode the left and right stereo channels. Stereo transmission includes a 19 kHz pilot tone and for correct decoding the phase of the reclaimed 19 kHz signal must suitably synchronise to that of the 19 kHz pilot tone.

High-quality audio equipment 183

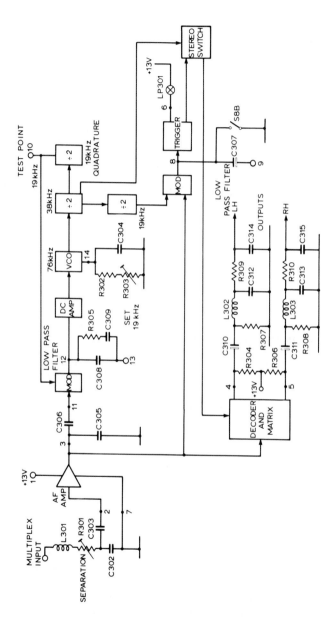

Fig. 7.8. Block diagram of the MC1310PQ stereo decoder i.c. and related discrete components as used in the circuit of Fig. 7.7. See text for description

## 19 kHz adjustment

To make sure that this requirement is achieved, the v.c.o. is adjustable by R303, and to Test Point 10, which delivers the reclaimed 19 kHz signal, can be connected a frequency counter, the preset then being adjusted until the signal is at exactly 19 kHz. The input to the stereo decoder should be muted for this adjustment. Unless the reclaimed signal is accurately phased with respect to the 19 kHz pilot tone stereo separation will be poor or non-existent.

A quick, though less accurate way of adjusting R303 is to tune to a stereo transmission and adjust the preset until the stereo beacon glows. The centre point between switch-on and switch-off of the beacon should be selected. If you have a stereo signal generator, then you could modulate one channel and monitor the other, adjusting the preset for the least breakthrough of modulation to the non-speaking channel. The BBC sometimes trasmits left and right test signals separately which could be used for adjusting both R303 and R301.

The required de-emphasis is achieved by L302/R307/C312/R309/C314 in the left channel and by the equivalent network in the right channel. The left and right outputs are then conveyed to the appropriate inputs of the control preamplifier section via the FM changeover switch.

General fault tracing should follow the pattern already established in Chapters 2, 3 and 4, but with high quality and hi-fi equipment care should be taken to use replacement components of corresponding values to the faulty originals. If possible, only transistors and i.c.s corresponding to the original types should be used. With early equipment, though, this may not always be possible, in which case a close equivalent should be sought. This is where semiconductor equivalent lists or handbooks can be useful. With regard to push-pull output stages, if a near equivalent has to be used to replace a transistor on one side, it is best to use a matching transistor for the other side. Also make sure that the class B quiescent current is set accurately to suit the circuit and devices used.

When there is no FM output in a circuit like Fig. 7.7 which uses a stereo decoder, you can check whether signal is being delivered to the decoder from the FM detector by connecting an earpiece to the multiplex input pin, which is pin 2 on IC4. If signal is present here but not on either of the stereo channels, then either the stereo decoder or an associated component would be at fault. Always remember, though, to make sure that the semiconductor devices are receiving supply power where appropriate.

If you have signal continuity through the decoder (though some circuits required 19 kHz pilot tone to switch on), you can often prove whether the signal 'break point' lies in the FM front-end or in the i.f./detector section by connecting an aerial to the input. Agreed, it is very unlikely that you would pick up an FM signal which would be accepted by the i.f. amplifier section, but owing to the high gain of the i.f. channel you can often hear a distorted something, if only at very low level. This, at least, proves that the fault lies in the front-end rather than in the i.f. channel and following circuits. If you have an FM signal generator, then the best thing would be to apply a 10.7 MHz signal to the i.f. input. You would (or should) certainly hear the modulation of this from the speaker!

To locate a point of signal discontinuity in the control preamplifier it may be possible to bypass a suspect stage with a capacitor, as told in Chapter 1 for testing suspect r.f. i.c. stages. With audio, though, you will need to use a capacitor around 0.1 µF. Too high a value should be avoided for fear of damaging a transistor or i.c. by charging current. You are not interested in the quality of the bass reproduction for quick tests of this kind.

With a circuit like that in Fig. 7.6(a), normal reproduction on tape replay but nothing on pickup or aux would strongly suggest trouble in the compound preamplifier VT601/VT602. With audio equipment in general there is usually not much of a problem in tracing the signal from input through the various stages using an earpiece or amplified signal detecting device. One of the biggest problems in rapid servicing is in locating the suspect stage or, at least, fault area. Once you have unearthed this it is not all that difficult to test around the area and components of the associated stage to bring to light the culprit. It is comforting to remember that the equipment *was* working before the fault occurred, though it must be admitted that with some equipment such a possibility can seem very remote, especially when trying to expose an obscure fault condition!

## Mains power supply

Equipment powered from the mains supply can produce hum symptoms. This results mainly, however, from deterioration of the reservoir or smoothing electrolytics or failure of the metal rectifiers. Bridge circuits are frequently employed owing to their greater efficiency over single or double rectifier circuits.

Overheating of the mains transformer should lead to a check of the supply current. If this is normal or low, the electrolytics or rectifier should be tested.

Some transistor equipment uses a regulated power supply. Here one or more transistors are used in a control circuit, with a zener diode providing a reference voltage. The function of the circuit is that the transistor collector passes current to the equipment, the amount of current that it passes being governed by the voltage across the load, since this sends a voltage back to the base either direct or through an amplifier. Thus, the base controls the supply current depending upon the load conditions (see the power supply in Fig. 7.6(a), for example).

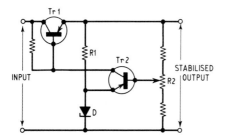

Fig. 7.9. Basic transistor feedback series stabiliser circuit. Feedback is via Tr2 to the base of Tr1, the emitter follower stabiliser

An example of a series transistor voltage stabilising circuit including a d.c. amplifier is shown in Fig. 7.9. Zener diode D provides a constant reference voltage, being biased into the zener breakdown region of its characteristic by R1. A part of the output voltage of the circuit is tapped off via R2 and applied to the base of the d.c. amplifier Tr2. The control voltage appearing at Tr2 collector is applied to the base of the series regulator transistor Tr1. This results in the collector-emitter voltage of Tr1 changing to oppose the initial variation in output.

**Fault Diagnosis Summary Chart 5: Radio Receivers and Amplifiers**

| Condition | Probable Cause | Check |
|---|---|---|
| Complete failure | (i) Exhausted battery or faulty power supply<br>(ii) Supply discontinuity in output stage<br>(iii) Disconnected or open-circuit speaker<br>(iv) Defective audio section<br>(v) Defective i.f. stages<br>(vi) Defective frequency changer stage<br>(vii) Misalignment | (i) Battery or power supply unit<br>(ii) Output transistors. Supply feed components. For switching 'thump' in speaker<br>(iii) Speaker. For switching 'thump' in speaker<br>(iv) 'Life' of stage by touching base of first a.f. transistor with screwdriver blade, etc. D.c. conditions of transistors<br>(v) I.f. stages for signal continuity and d.c. conditions<br>(vi) For oscillations with second set. Transistor. D.c. conditions<br>(vii) Alignment of all stages |
| Distortion | (i) Low supply voltage<br>(ii) Incorrect audio biasing<br>(iii) Leaky or low value a.f. coupler<br>(iv) Defective a.f. transistor<br>(v) Overloading | (i) Battery or supply voltage<br>(ii) For crossover distortion. Base and emitter resistors in audio stages<br>(iii) Coupling capacitors<br>(iv) A.f. transistors<br>(v) A.g.c. system |
| Instability | (i) High resistance battery<br>(ii) Faulty decoupling<br>(iii) Misalignment | (i) Battery<br>(ii) Decoupling capacitors<br>(iii) Alignment and neutralising |
| Motor-Boating | (i) Defective supply filter capacitor<br>(ii) Defective a.f. coupler<br>(iii) Exhausted battery | (i) Electrolytic capacitors on supply circuit<br>(ii) A.f. coupling capacitors<br>(iii) Battery |

**Fault Diagnosis Summary Chart 5: Continued**

| Condition | Probable Cause | Check |
|---|---|---|
| Lack of sensitivity | (i) Misalignment<br>(ii) Low supply voltage<br>(iii) Fracture of ferrite aerial<br>(vi) Incorrect bias on i.f. transistor<br><br>(v) Defective i.f. transformer<br>(vi) Low gain a.f. channel<br>(vii) Open-circuit emitter capacitor<br>(viii) Open-circuit base bypass capacitor | (i) Alignment of all stages<br>(ii) Battery or supply voltage<br>(iii) Ferrite rod mechanically<br>(iv) Base resistors (and possibly emitter resistors) on i.f. transistors. A.g.c. bias<br>(v) I.f. transformers and fixed capacitors across windings<br>(vi) A.f. channel gain (see Chapter 4)<br>(vii) Emitter capacitors in all stages<br>(viii) Base bypass capacitors in all stages |
| Intermittent operation | (i) Hair-line fracture in printed circuit<br><br>(ii) Exhausted battery<br>(iii) Dry joint<br><br>(iv) Defective speaker | (i) Printed circuit by applying localised pressure to the various sections in turn (see Chapter 8)<br>(ii) Battery<br>(iii) Soldered connections on interconnecting leads and printed circuit<br>(iv) Speaker and connecting leads, plugs, sockets and etc. |

# 8 Practice of transistor equipment servicing

The servicing of transistor equipment, like any other kind of equipment for that matter, resolves to two major operations. The first is *the finding* of the fault and the second is *the making* of the actual repair. While past chapters have shown ways by which troubles in the circuits can be quickly found, this chapter deals mainly with the practice of servicing.

For serious work, four main items of test equipment are desirable. These are: (i) multirange testmeter; (ii) audio-frequency generator; (iii) radio-frequency generator AM and FM; and (iv) oscilloscope.

The multirange testmeter will include sections scaled as (a) voltmeter, (b) current meter and (c) ohmmeter. Each section will be switchable over various ranges. Let us consider each section in turn, starting with the voltmeter.

## Multirange testmeter

For transistor work, the voltmeter section should have a sensitivity of not less than 20 000 ohms/volt and there should be at least three ranges giving full-scale deflections of 1000 V, 10 V and 1 V. A four-range section of 250 V, 25 V, 2.5 V and 1 V or less f.s.d. would be even better. In any event, a very low voltage first range is essential bearing in mind the frequent need to measure voltages of less than 1, often in the order of 250 mV.

The current section is not so important as the voltmeter section, for low currents can be measured, as we have seen (Chapter 2), by measuring the voltage they develop across feed resistors of known value and then bringing in Ohm's law. Nevertheless, the current consumption of a complete piece or section of equipment is often useful to know, so a current range up to about 100 mA f.s.d. can be well accommodated. In practice, of course, multirange testmeters have the lowest current range based on the movement sensitivity. A movement providing a voltmeter sensitivity of 20 000 ohms/volt has a full-scale current deflection of 50 µA. This will probably be shunted on the lowest current range (and on the higher ranges, of course) to give a f.s.d. of probably 0.1 mA, though some instruments may exploit the full sensitivity of the movement on the lowest current range.

Although not essential, a.c. voltage and current measuring facilities can be useful when mains-operated equipment is encountered. An a.c. voltmeter can also be employed to measure audio power across a load resistor of known value in accordance with the expression $W = E^2/R$, where $W$ is the power in watts, $E$ the a.c. voltage across the load $R$. $R$ can be the existing load in the equipment, such as a loudspeaker, or an artifical load purposely introduced. The latter is desirable.

The ohmmeter section should be powered internally from a battery of no larger than 1.5 V on one range. A greater voltage could incite forward currents in transistor and semiconductor junctions during a test of strength sufficient to cause their failure or to cause a change in their characteristics.

Ranges of 1 000-ohm, 10 000-ohm, 100 000-ohm and 1 000 000-ohm are useful, though in practice such a wide range is not essential. It is a good idea to be able to assess the operation of a transistor junction by checking its forward and reverse resistances with an ohmmeter. It has already been mentioned in Chapter 2 that, when measuring the value of in situ resistors in a transistor circuit, forward conduction of a transistor junction can well produce very misleading results. If there is any doubt about the accuracy of a resistance reading, the measurement should be made a second time with the ohmmeter leads reversed so as to change the direction of current flow through the junction.

## A.F. generator

This instrument is essential for audio frequency response measurements of hi-fi amplifiers. It is also useful for stage-to-stage signal continuity tests in any audio equipment. For this application, however, an audio generator of a fixed frequency (usually 400 Hz or

1 000 Hz) as embodied in a modulated r.f. signal generator is adequate. Signal injector instruments, employing a multivibrator type of oscillator, are also available in self-contained portable form for signal continuity tests both in r.f. and audio equipment. The repetition frequency is arranged to be in the audio range, and harmonics of the signal spread well into the r.f. spectrum, giving the r.f. facility as well.

Signal injectors are ideal for speedy, point-to-point signal tests, but a proper a.f. generator is demanded for tests of a more serious nature. The generator proper features a variable attenuator feeding (usually) into 600 ohms, and the frequency range often extends from a few cycles per second up to low r.f. Some instruments also have facilities for switching from a very pure sine wave to a good quality square wave.

Unless feeding into a matching impedance, the generator signal should be applied via an isolating capacitor (about 0.1 µF) and a resistor of about 10 000 ohms. A very high signal amplitude should not be used in low-level stages as this could damage the transistors.

### R.F. generator

The frequency range of the r.f. generator will be governed by the nature of the equipment under test. As already mentioned, there will be an inbuilt audio oscillator for modulating the carrier wave. The modulation frequency is generally fixed, as also is the depth of modulation (to about 30 per cent), but on some instruments the modulation depth is adjustable.

The r.f. signal is developed across a relatively low impedance (about 75 ohms) and this is coupled from the r.f. oscillator via an attenuator system, often calibrated in terms of millivolts and microvolts (or in decibels).

The signal can be connected direct to the input (say the base) of an r.f. transistor via an isolating capacitor of about 0.1 µF. However, direct coupling should be avoided where the generator is likely to load the circuit receiver. Here the generator signal should be injected into the ferrite rod via a completely isolated coupling loop, as described in Chapter 7. Again, too great a signal level should be avoided. If there is no response at moderate signal levels, then something is wrong with the circuit and there is no point in turning up the input signal. This will only force the signal through the defective section and probably damage a transistor in the process.

For FM work, the r.f. generator should be switchable to FM and tunable over the FM band (80–108 MHz) and around 10.7 MHz for i.f. tests. An output impedance of 75 ohms unbalanced is useful. The more detailed FM tests and instrument applications are given in *Audio Equipment Tests* (Newnes Technical Books).

## Oscilloscope

The oscilloscope is useful as a signal tracer to check the signal waveform at various points along a circuit when a signal is applied from a generator at the input. Visual display of the waveform reveals whether or not a stage is being overdriven due to a too strong input signal. Overloading shows up as clipping at the top or bottom (or both) of the waveform display.

Most oscilloscopes respond on the Y channel to audio signals and low r.f. signals, but only the more expensive versions give a response to higher frequencies up to 10 MHz or more. The Y input of an oscilloscope is usually of very high impedance, so it can be applied to any part of the circuit to respond to signal *voltage* without damping the circuit in any way.

It is often necessary to monitor current waveforms in transistor equipment, since the transistor is a *current* operated device, and this can be accomplished by introducing a low value resistor in series with the signal current and then monitoring the voltage produced across that resistor. The trace will then be that of a current waveform. The idea is shown in Fig. 8.1.

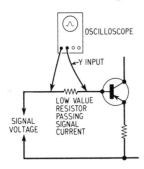

Fig. 8.1. A current waveform can be displayed on an oscilloscope by introducing a low value resistor in series with the circuit passing the signal current and then applying the voltage developed across this to the Y input of the oscilloscope as shown here. The resistor should not have too great a value to disturb the normal operation of the circuit

The majority of transistor equipment faults can be speedily located aided by the equipment mentioned in the foregoing. Many faults, however, can be diagnosed with virtually no equipment at all, as revealed in past chapters. Sometimes, though, extra special test equipment is demanded, particularly where the equipment itself is of a specialised nature.

## Electronic voltmeter

A electronic voltmeter can sometimes aid fault diagnosis, especially in cases where an ordinary multirange testmeter would impose too great a load on the circuit under test. An electronic voltmeter has a very high terminal resistance or impedance, often in the order of megohms. Such instruments, however, have the disadvantage of having to be powered from the mains supply or battery. Mains-powered instruments can give rise to certain complications when used with transistor equipment.

## Audio millivoltmeter

An audio millivoltmeter is another valuable test instrument for audio work. Although called a 'millivoltmeter' because it measures very small audio (a.c.) voltages, sometimes down to $10\,\mu V$, from a few Hz and right up to 100 kHz or more, it can be switched to measure much higher voltages, sometimes up to 300 V or more. It is an essential instrument for hi-fi amplifier testing.

## Signal tracer

Another useful instrument is a signal tracer. In simplest form this is just a pair of headphones or an earpiece, the idea being to connect it across the signal path at various points in the circuit so as to discover, for instance, at which point the signal disappears. A tracer can also be useful for locating the origin of distortion.

A more elaborate signal tracer is one which incorporates its own transistor a.f. amplifier. The output of the amplifier is connected to a small speaker or earpiece, and the amplifier, of course, makes it possible to detect signals of a much smaller amplitude than by the earpiece alone. Transistorised signal tracers and partnering multivibrator type generators are available commercially.

## Transient problems

Inadvertent, high amplitude currents, even of a transitory nature, can under certain conditions flow through a transistor junction and cause its destruction. Such destroying currents are sometimes encouraged when mains powered test equipment is used to locate faults in transistor equipment. For this reason isolation of some kind should always be used between the instrument and the equipment under test.

Capacitor isolation by itself is not always sufficient to prevent transient damage, especially if the mains-powered instrument is connected and disconnected while it is actually connected to the mains supply and switched on. Some instruments have an r.f. filter circuit on the mains supply and this can sometimes pass a little mains current to the metal parts of the instrument. The metal is usually the 'common' earth connection, and unless the instrument is properly earthed mains currents can get into the transistor junctions. It always pays therefore, to make sure that all mains-powered test equipment is adequately bonded to a good earth point before it is connected to the equipment under test.

Electrostatic energy from the mains supply can also be transferred from an unearthed instrument to a transistor junction via capacitance between the primary and secondary windings of the instrument's mains transformer or via an isolating capacitor connected between the instrument and the circuit under test.

Now, if the earthed instrument is connected to the equipment under test and an unearthed soldering iron is employed, say to make a disconnection at the base of a transistor in the equipment, electrostatic energy can again flow due to capacitance from the element in the iron, via the transistor emitter junction, back to earth. This will almost certainly ruin the transistor. Moreover, soldering irons tend to develop small leakages between the element and metal bit, not sufficient to be harmful to a person but adequate to cause the flow of high amplitude transient currents. When dealing with 'delicate' MOSFET devices a very strict procedure is demanded to avoid damage due to transients or static.

Other ways by which transistors can suffer transient damage include the shorting of transistor lead-outs, the replacing of transistors or components and the soldering of any part of the circuit while the equipment is powered.

Damage may also result if the lead-outs are bent closer than 1.5 mm from the seal, if an ohmmeter with an internal battery greater than 1.5 volts is used for continuity checks, if the power supply or battery is connected to the equipment in reverse of its correct polarity, if the lead-outs of a transistor are incorrectly connected to the circuit and if the light-proofing on the case is damaged.

### Heat shunt

Damage can also occur due to the transference of excessive heat from a soldering iron, via the lead-out wires, to the junctions. This can be avoided by the use of a heat shunt clamped on the lead-out wire when the end of it is being soldered or unsoldered. A heat shunt is nothing more than a bulk of metal which blocks heat conduction along the leadouts. A pair of pointed-nose pliers makes a good heat shunt, but is not very convenient to use.

While heat shunts are available commercially, they can be made quite easily from crocodile clips. The idea is to fill the jaws of the clip with solder to form two solid blocks. The clip is then clamped to the lead-out wire being soldered so that the two blocks of solder press tightly against the wire.

## Printed-circuit board repairs

Most transistor equipment is built upon printed-circuit boards. These are laminated, plastic sheets, which are good electrical insulators, carrying the 'printed wiring'. The boards start life as copper-foil-clad plastic sheets and a chemical process is employed to etch away the copper-foil not needed, thereby leaving the circuit traced in copper upon the insulant. The foil is very thin, about $0.07$ mm ($0.003$ in) and can thus suffer damage from rough treatment.

Small holes are drilled in the circuit board at the correct points in the printed circuit to accommodate the lead-out wires of the components, and many boards have the reference numbers of the components printed against them so that they may quickly be identified in the circuit diagram.

Unfortunately, it happens – not infrequently – that a component leadout wire is not successfully soldered to the copper circuit, a dry joint thus being present. Such trouble, of course, can result in crackles and intermittent operation, and it has been known for poor electrical connections of this kind to impair the efficiency of a circuit transistor.

## Printed wiring faults

Another cause of trouble is a hair-line break somewhere in the printed wiring. It may be impossible to see the circuit break with the naked-eye. A small magnifying glass is useful for locating such trouble. Applying strain or stress to the printed-circuit board may break or make the connection at the fracture. Care should be taken, however, to avoid undue stress being applied to the board, as extreme flexing can cause the fracture of other parts of the circuit.

Dry joints and fractured printed wiring can often be located by gently tapping the board with the handle of a screwdriver at various points over its surface while listening at the speaker (in the case of a radio set, amplifier, and etc.) with the volume control only moderately advanced. When the section of the circuit which is at fault is so disturbed a distinct crackle is often produced in the speaker.

By tapping over the circuit area and by varying the strength of the tapping it is often possible to localise the defective connection to a very small area. It may also be necessary gently to tap the components. Hard tapping should be avoided.

When the break is eventually located it can be repaired by the application of a blob of solder or, if necessary, by soldering a piece of wire between the two halves of the conducting channel.

**Soldering**

Excessive heat should be avoided on printed-circuit boards, as the adhesive holding the copper circuit to the board may be melted. Too much heat from a soldering iron can also damage nearby components. The best plan is to employ a soldering iron with a small, clean and well tinned bit having a loading in the order of 25 watts. The soldering should be done as quickly as possible, and only resin-cored solder of the kind designed for printed circuit applications should be used. This has a lower melting temperature than solders designed for ordinary wiring. On no account should acid fluxes be used.

The circuit on a printed board can sometimes be traced relative to the components, the references of which are on the top of the board (i.e., the opposite side to the foil circuit), by placing a lit electric light bulb below the board. The light from this causes the circuit to show as shadows on the top of the board. The use of a magnifying glass will then reveal very fine circuit fractures.

After soldering the components to a printed circuit, some makers coat the circuit with an insulating varnish. This must, of course, be removed before any soldering on the circuit is attempted or before voltage measurements are made by applying the point of a test probe to the circuit sections. A good solvent for the varnish is acetone (the liquid with the smell of pear drops). This should be applied only to the section of the circuit under attention with a soft lint-free cloth. Only a very small amount of the liquid should be used.

**How to extract a faulty component and fit a replacement**

The best way to extract a defective component from a printed-circuit board is to cut the lead-out wires with a small pair of sharp side cutters. The wires may be removed by applying the tip of a hot soldering iron to the soldered joint beneath the board, letting the old wires fall out, and inserting the lead-out wires of the new component through the now vacant holes in the board.

Sometimes it is difficult to get the old component wires out of the circuit without overheating the board, with the possibility of damaging it and adjacent components. In this event, the replacement component can be soldered direct to the lead-out wires of the old component.

Some components, such as coils, transformers, presets, potentiometers and the like, are fixed to the board by tags. These are generally located in slits in the board, twisted slightly for mechanical rigidity and soldered. This kind of component, again, is best removed by cutting the tags close to the body of the component and then applying heat to the solder around the tag beneath the board, wriggling the tag with the tip of the iron, hoping that it will drop from the board. If it fails to do this, a small pair of tweezers should be used to extract it while the solder around it is still molten.

Extra special care should be taken when a component is removed for test only or for the substitution of a test component. Here the lead-out wires or tags must be removed without cutting and a heat shunt should be clamped on the lead-out wires.

When soldering in new components, especially semiconductor diodes and transistors, a heat shunt must always be used. It is desirable to remove this sort of component from the board for test *only* as a last resort.

To finalise a printed-circuit board repair, the circuit side of the board should be scrutinised to ensure that no splashes of solder remain on the conductors or, indeed, between adjacent conductors. Only when one is perfectly satisfied on this point should the equipment be powered. Remember that an inter-conductor short can easily result in the need to replace a costly i.c., transistor or other component.

Tools required for the repair of printed-circuit board equipment include several sizes of good, sharp side cutters, pointed nose pliers, tweezers of various sizes, two or three medium- and low-wattage soldering irons with different size bits plus the usual screwdrivers and pliers.

## Transistor and i.c. precautions

Transistors and i.c.s are inherently robust devices. They can, however, easily be damaged if mishandled or misused. It is therefore worthwhile summarising the following points:

1. *Don't short-circuit lead-out wires.* This can easily happen when using a screwdriver, test probe, etc.
2. *Don't carry out soldering operations with the equipment switched on.* Short-circuits and surges can easily occur in these circumstances.
3. *Don't use an unearthed soldering iron.* The insulation between the element and bit of the iron may break down so that the bit becomes 'live'.
4. *Don't change components with the equipment switched on.* Surges large enough to destroy transistors can be created by doing this.
5. *Don't use a soldering iron without a heat shunt.* Excessive heat can damage transistors and miniature components.
6. *Don't connect batteries incorrectly.* Reversing the polarity of the supply can permanently affect transistor characteristics.
7. *Don't make resistance measurements or continuity tests with an ohmmeter giving an output voltage greater than 1.5 V.*
8. *Don't connect transistors in the circuit the wrong way round.* The transistor characteristics may be permanently altered.
9. *Don't damage the light-proofing coat on a transistor.* Light can affect the currents flowing in a transistor.
10. *Don't bend transistor lead-outs nearer than 1.5 mm from the seal.* Otherwise the seal may be damaged.

## Transistor parameters

While complete standardisation still fails to exist in transistor terminology, the following list of the terms used in some texts and adopted by some device manufacturers, along with the definitions, may prove useful. Unless othewise stated, it should be assumed that they are applicable to static (d.c.) conditions of test.

$BV_{CEO}$    Collector/emitter breakdown voltage, base open circuit.

$BV_{CER}$    Collector/emitter breakdown voltage with specified resistance between base/emitter.

$BV_{CBO}$    Collector/base breakdown voltage, emitter open circuit.

Transistor perameters 199

| | |
|---|---|
| $BV_{CERL}$ | Collector/emitter breakdown voltage with specified resistance between base/emitter and with a specified resistance in the collector circuit. |
| $BV_{CES}$ | Collector/emitter breakdown voltage, base connected to emitter. |
| $BV_{CEV}$ | Collector/emitter breakdown voltage with base reverse-biased with respect to emitter. |
| $BV_{CEX}$ | Collector/emitter breakdown voltage with base reverse-biased with respect to emitter through specified circuit or under specified conditions. |
| $BV_{EBO}$ | Emitter/base breakdown voltage, collector open circuit. |
| $C_{ib}$ | Common-base emitter/base input capacitance. |
| $C_{ie}$ | Common-emitter base/emitter input capacitance. |
| $C_{ob}$ | Common-base collector/base output capacitance. |
| $C_{eo}$ | Common-emitter collector/emitter output capacitance. |
| $f_{hfe}$ | Small-signal common-base forward-current transfer-ratio cut-off frequency. |
| $f_T$ | Gain/bandwidth product, common-emitter mode. |
| $h_{FB}$ | Common-base forward-current-transfer ratio. |
| $h_{fb}$ | Small-signal common-base forward-current transfer ratio. |
| $h_{FE}$ | Common-emitter forward-current transfer ratio. |
| $h_{fe}$ | Small-signal common-emitter forward-current transfer ratio. |
| $h_{RB}$ | Common-base open-circuit reverse-voltage transfer ratio. |
| $h_{RE}$ | Common-emitter open-circuit reverse-voltage transfer ratio. |
| $I_B$ | Base current. |
| $I_C$ | Collector current. |
| $I_{CBO}$ | Collector-cut-off current, emitter open. |
| $I_{CEO}$ | Collector-cut-off current, base open. |
| $I_E$ | Emitter current. |
| $I_{EBO}$ | Emitter-cut-off current, collector open. |
| $P_C$ | Collector power dissipation. |
| $P_T$ | Total transistor dissipation. |
| $Q_{SB}$ | Stored base charge. |
| $V_{BC}$ | Base/collector voltage. |
| $V_{BE}$ | Base/emitter voltage. |
| $V_{CB}$ | Collector/base voltage. |
| $V_{CE}$ | Collector/emitter voltage. |
| $V_{EB}$ | Emitter/base voltage. |
| $V_{EC}$ | Emitter/collector voltage. |
| $V_{RT}$ | Punch-through (reach-through) voltage. |

# Index

Acceptors, 7
Aerial amplifiers for television, 83, 124–125
Aerial circuit alignment, 163
A.f. generator, 190–191
A.g.c., 130, 138
Alignment,
  a.m., 161
  f.m., 164–171
  oscillator, 162
Astable circuits, 150–153
Audio amplifier arrangements, 92
Audio amplifier circuit, two-stage, 61
Audio amplifier, complete failure, 95
Audio and video amplifiers: Fault Diagnosis Summary Chart 2, 118
Audio and video circuits, fault-finding, 92–118
Audio circuit faults, 94
Avalanche breakdown voltage, 9
Avalanche diode, 29

Base, 12
Basic circuit test: Fault Diagnosis Summary Chart 1, 74
Beta test, 57
Bias oscillator circuits, tape recorder, 144
Bias stabilising diode, 100
Bistable circuits, 153
Blocking oscillator, 150

Blocking oscillator faults, 151
Bottomed oscillator, 142
Bottoming, 75, 143
Breakdown voltage, 9

Capacitive reactance, 98, 110
Capacitor diode, 26–28
Capacitor leakage, 62
Capacitor values and polarisation, 55
Checking beta, 57
Checking current without breaking circuit, 56
Checking gain, 133
Checking total current, 51
Circuit analysis, 56
Class A operation, 52
Class B operation, 53
Class C operation, 53
Clipping, 75
Collector, 12
Collector-emitter feedback, 143
Collector junction, 12
Colour receivers, 116–117
Colpitts circuit, modified, 125
Common-base circuit, 18, 20–21
Common-emitter circuit, 18–19, 20, 77
Complementary audio amplifiers, 65–68, 107–109
Complementary pair, 66
Component replacement, 196

Conventional current flow, 2–3
Coupling capacitor, leaky, 62
Cross modulation, 137
Crossover distortion, 99
Current carriers, 2
Current gain, 22, 78
Current tests, 56

Darlington pair, 25
D.c. conditions in complementary circuits, 60
Degenerative feedback, 111
Depletion f.e.t., 31, 35
Detector and a.g.c. stage, 145
Diac, 37
Diode characteristics, 8–9
Diode effect, 6
Directly coupled circuits, 63
Distortion, 99, 105, 106, 173
Donors, 7
Drain, f.e.t., 30
Drift field, 16
Dual-gate f.e.t., 31–32

Eccles-Jordan circuit, 153
Electron flow, 3
Electron vacancies, 4
Emitter, 12
Emitter, junction, 12
Enhancement f.e.t., 31, 34, 35
Epitaxial planar transistor, 16
Equalisation,
    f.m., 104
    PU, 101
    servicing, 102
    tape, 104

Failure, audio amplifier, 95
Failure, power supply, 42
Failure, transistor, 50–51
Fault Diagnosis Summary Chart 1: Basic circuit tests, 74
Fault Diagnosis Summary Chart 2: Audio and video amplifiers, 118
Fault Diagnosis Summary Chart 3: R.f. amplifiers, 139
Fault Diagnosis Summary Chart 4: Oscillators, 155
Fault Diagnosis Summary Chart 5: Radio receivers and amplifiers, 186–187

Fault-finding:
    audio and video circuit, 92–108
    oscillator stages, 141–155
    r.f. amplifier circuit, 119, 132
    transistor radio and hi-fi, amplifier, 156–186
Faulty component – how to extract and fit a replacement, 196
Ferrite beads, 137
Ferrite rod aerial, 121
Field effect transistors, 30–31
    circuits, 32–33,
    testing, 73
'Flip-flop' circuit, 153
Forward characteristic, 9
Forward conduction, 6
Forward current direction, 12
Forward gain control, 13, 130
F.m. receiver using i.c., 181
Free electrons, 2
Frequency response, 97
    h.f., 109

Gain, 89, 133–135
Gain-bandwidth product, 79–80
Gain control, 138
    forward system, 130
    reverse system, 130
Gate, f.e.t., 30

Half-wave line, 128
Heat shunt, 194
Heat sinks, 17, 111–112
Hi-fi audio equipment, 175–186
Holes, 3–4
$h$ parameters, 80
Hum test, 96
Hybrid circuits, 48

I.f. alignment, 162, 165
I.f. i.c. circuit, 43
Impedance considerations, 85
Impurity, 3
Input and output impedance, 24
Input charcteristic, 21
Instability, 111
Instability in m.f. amplifiers, 135
Instability in v.h.f. and u.h.f. amplifiers, 136

Insulated gate f.e.t., 31
Integrated circuits, 37
   TV sound channel, 39,
   CA 3043, 41
   digital, 45
   testing, 70–73
Internal impedances, 88
Inter-stage losses, 93

Junction diode, 4–5
Junction f.e.t., 33
Junction transistor, 12

Knee voltage, 24

Lack of oscillation, 147
Lack of response, 133
Leakage current, 25
Leaky coupler check, 62
Light-emitting diode, 29–30
Load line, 78–79
Local oscillator check, 172
Low gain, 134
Low-level distortion, 105

Main $h$ parameters and equivalents, 80
Mains power supply, 185
Matching, 90
Measuring distortion, 106–107
Medium-frequency stages, 121
Mesa transistor, 16
MOSFET, 31
MOSFET v.h.f. mixer, 32
Microcircuits, 48
Microphone amplifier circuit, 64
Minority carriers, 7–8
Misalignment, 173
Mixer diode, 29
Mobile electrons, 3
Modified Colpitts circuit, 144
Modulation, 148
Monitor current waveforms, 192
Monostable circuits, 153
Multirange testmeter, 189
Multi-stage equipment, 60
Multivibrators, 151–153

Negative current carriers, 3
Negative feedback, 64, 177

Neutralisation, 123
Noise, 104
Non-linear devices, 157
N-p-n transistors, 12
N-type semiconductor, 4

Ohmmeters, circuit and use of, 69
Op-amps, 44
Open-circuit collector junction, 58
Open-circuit emitter junction, 59
Operating conditions, transistors, 53–55
Oscillation, lack of, 147
Oscillator alignment, 162
Oscillator, push-pull, 132, 145
Oscillator stages, fault-finding in,
   141–155
   Fault Diagnosis Summary Chart 4, 155
Oscillator tests, 146
Oscilloscope alignment, 166, 192
Output characteristics, 24
Output impedance, 24
Overloading, 137, 174

Phase-shift oscillator, 153
Photoelectric current, 8
Pilot-tone adjustment, 189
P-i-n diode, 29
Planar transistors, 16
P-n junction characteristic curve, 10–11
P-n-p transistors, 12
Point-contact diode, 7
Point-to-point signal tests, 193
Positive current carriers, 4
Positive holes, 3
Potential barrier, 4–5
Power amplifier, 131–132
   a.f., 177
Power gain, 23, 78
Power supply, 160
   failure, 51
   test, 51–52
Preliminary circuit and transistor tests,
   50–74
Printed-circuit board repairs, 195
Printed wiring faults, 195
P-type semiconductors, 4–5
Push-pull amplifiers, 94, 108, 176
Push-pull effect, 67
Push-pull oscillator, 145

## 204 Index

Quarter-wave line, 128
Quiescent current, setting, 179

Radio receiver, transistor, 122, 157–160
Radio receivers and amplifiers: Fault Diagnosis Summary Chart 5, 186–187
Read diode, 29
Recombination, 7
Reference diodes, 10
Resonant line u.h.f. amplifier, 127–129
Reverse characteristic, diode, 9
Reverse conduction, 6
Reverse system of gain control, 130
R.f. amplifier circuits, fault-finding in, 119–139
R.f. amplifiers, Fault diagnosis Summary Chart 3, 139
R.f. generator, 191–192

Self-oscillating mixer technique, 156
Semiconductor, 2
Semiconductor junction, 4–6
Sensitivity specification, 77
Series-connected tuned circuits, 124
Series stabiliser circuit, 188
Servicing hints, i.c.s, 46, 189–197
Shorting collector junction, 58
Signal conditions and tests, 75–91
Signal effects, 79
Signal generator, 190–192
Signal generator matching, 85
Signal injection, 82
  testing, 83
Signal injectors, 82–83
Signal tests, 83
Signal tracer, 95, 193
Signal tracing, 84
Signal, types of, 81, 119
Soldering, 196
Source, f.e.t., 30
Spurious oscillation, 60
Spurious signals, 136–137
Squegging, 147
Stabilising d.c. feedback, 67
Stage gain, basic, 76
Stagger tuning, 123–124
Standard British a.m. broadcast i.f., 162
Static characteristics, 21
Schottky diode, 29

Stereo decoder, 180, 183
Stereo separation, 182
Symptoms:
  bottoming, 75, 143
  clipping, 75
  complete failure of audio amplifier, 95
  crossover distortion, 99
  cross modulation, 137
  distortion, 99, 105, 173
  hum, 96
  instability, 111
  instability in m.f. amplifiers, 135
  instability in v.h.f. and u.h.f. amplifiers, 136
  lack of oscillation, 147
  lack of response, 133
  low gain, 134
  low level distortion, 105
  misalignment, 173
  noise, 104
  overloading, 137, 174
  poor frequency response, 97, 109
  spurious oscillation, 60
  spurious signals, 60
  squegging, 147
  thermal runaway, 19

Television i.c. sound channel, 39
Testmeter, 68
Thermal runaway, 19–20
Thyristor, 36
  motor control, 37
Tone control system, 100, 101, 103, 104
Transfer characteristic, 22
Transient problems, 193
Transistor,
  alloy-diffused, 15
  circuit modes, 18–21
  effect, 5–6
  epitaxial planar, 16
  failure, 50–51
  fundamentals, 1–49
  mesa, 16
  parameters, 198
  precautions, 198–199
  types, basic, 15–18
  unijunction, 36
Transistor radios and hi-fi amplifiers – fault-finding, 156–188
Transition frequency, 79–80
Transmission line, 128

Triac, 37
Tuned oscillators, 141
Tunnel diode, 28–29
Trappett diode, 29
Turnover voltage, 8

U.h.f. amplifiers, 127, 129
U.h.f. effect of small inductance at, 120
U.h.f. tuner, 27, 129–130
Unilateralisation, 123
Unijunction transistor, 36

Valve voltmeter, 193
Varactor diode, 26–28
Varicap, 26
V.h.f. aerial amplifiers, 126

Video circuits, 113
Video-frequency amplifiers, 93
Voltage gain, 23, 78
Voltage regulator diodes, 10

Wattmeter, 161
Wideband amplifier, 125–126
Wien network oscillator, 154

$Y$ parameters, 80

Zener current, 10
Zener diodes, 10, 188
Zener effect, 9
Zener voltage, 9
$z$ parameters, 81

## Video Techniques
### Second edition
### Gordon White

A comprehensive treatment of the many aspects of video written for the engineer or technician working in television or ancillary industries. The book will also be useful to anyone seeking information outside his or her own particular specialization and to production staff and electronics engineers who need a working knowledge of techniques in this rapidly changing industry.

This second edition has been updated throughout, and enlarged, with new information on satellite broadcasting, standards conversion, CCD camera tubes, ENG cameras and recorders, interactive video, conference television, fibre-optics, world standards, FST and FSQ picture tubes, and so on.

0 434 92290 0
336pp 216 × 138mm 100 diagrams 100 photographs
Hardback

## Video Handbook
### Second edition
### Ru Van Wezel

The *Video Handbook* presents in one volume most of the information needed to understand video. It is a practical book on all aspects of the subject intended for the serious enthusiast, the student and the semi-professional in video as well as technical personnel in small video/television production companies. Since the first edition was sold out all the cameras and video recorders have been replaced by one or more generations and almost every circuit has been modified, improved or replaced. The book has, therefore, been virtually re-written and entirely re-set.

0 434 92189 0
432pp 234 × 156mm 120 diagrams 44 photographs
Hardback

## Newnes Television and Video Engineer's Pocket Book
### Eugene Trundle

Along with 'formal' reference material such as channel allocations, world-wide TV systems and videorecorder format specifications, the author gives such practical advice as how to set-up a colour picture tube, how to service a video deck, and details of conversion of foreign equipment for local use.

Without neglecting the basic theory, the emphasis is on modern equipment: frequency-synthesis tuning, satellite-dish head-ends, single tube colour cameras, Video 8 tape format, depth-multiplex sound and digital servo systems. A long chapter describes test instruments, fault-finding and repair techniques, illustrating points made with off-screen photos.

0 434 90197 0
280pp 190 × 110mm 155 diagrams 12 photographs
Hardback

## Domestic Videocassette Recorders
### A servicing guide
### Second edition
### Steve Beeching

'Those who want a practical book on VCR techniques will find this work an essential guide and reference.'
*Television*, on first edition

0 434 90121 0
Paperback

# Colour Television Servicing
## Second edition
### Gordon J. King

0 408 00464 9
Paperback

# Electronic Circuits Handbook
## Design, testing and construction
### Michael Tooley

This book provides the reader with a unique collection of practical working circuits together with supporting information. Circuits can be produced in the shortest possible time and without recourse to theoretical texts.

All the circuits described have been thoroughly tested and a range of commonly available low-cost components has been adopted. The simplest and most cost-effective solution is adopted in each case.

Furthermore, the circuits can readily be modified and extended by readers to meet their own individual needs. Related circuits have been grouped together and cross-referenced within the text (and also in the index) so that readers are aware of which circuits can be readily connected together to form more complex systems. As far as possible, a common range of supply voltages, signal levels and impedances has been adopted. Ten test gear projects have also been included.

The book is aimed at: practising industrial electronic technicians and engineers needing a quick reference guide to circuit design; technicians and engineers (in various fields) who wish to update their knowledge of electronics and requiring a practical introduction to the subject; students at GCSE, O-level, A-level, BTEC and City and Guilds levels requiring practical information related to project work.

0 434 91968 3
300pp 216 × 138 250 diagrams and 20 photographs
Hardback

# Servicing Digital Circuits in TV Receivers
### R. Fisher

'A clearly-written, well-illustrated theoretical text book.' *Radio & Electronics World*

0 408 01149 1
Paperback

# Electronics for Electricians and Engineers
### Ian Sinclair

The rapidly changing technology of electronics has left many technicians requiring an urgent updating of their skills. The same problems have faced other engineers, who now require a knowledge of electronics to understand new developments in their own subjects. Hardly any part of modern life is untouched by electronics, and almost everyone working in engineering or science needs some understanding of electronics.

This book has been written in response to the need that now undoubtedly exists. The aim has been to explain principles and devices in clear terms, without assuming a high level of prior knowledge. The reader will probably have some elementary knowledge of electricity, more likely from a practical than a theoretical standpoint. The early chapters of the book are concerned with a revision of modern electrical principles. In the sections that deal with electronics, the emphasis has been on principles and devices, but sufficient circuit diagrams have been included to illustrate how the various electronic components are used.

The book provides a practical introduction to electronics for the electrician or newly qualified technician and a welcome retraining course for the non-electrical engineer.

0 434 91837 7
288pp 216 × 138mm 200 diagrams and 10 photographs
Paperback